AF556097

Dairy Chemistry and Animal Nutrition

Dairy Chemistry and Animal Nutrition

Umesh Kumar

Dairy Chemistry and Animal Nutrition

ISBN 978-93-5111-376-8

Published in 2014 in India by

Reprint 2020

RANDOM PUBLICATIONS

4376-A/4B, Gali Murari Lal, Ansari Road
New Delhi-110 002
Phone : +91-11-43580356, +91-11-23289044
e-mail: randomexports@gmail.com, sales@randompublications.com,
info@randompublications.com

REPRINT-2025

Type Setting by : Keystoneprintads, Delhi-110051
Printed at : Mehra Printers, Delhi-110 092

Preface

Dairy cattle are cattle cows bred for the ability to produce large quantities of milk, from which dairy products are made. Dairy cows generally are of the species Bos taurus. Historically, there was little distinction between dairy cattle and beef cattle, with the same stock often being used for both meat and milk production. Today, the bovine industry is more specialized and most dairy cattle have been bred to produce large volumes of milk. The United States dairy herd produced 83.9 billion kg of milk in 2007, up from 52.6 billion kg in 1950. Yet there are more than 9 million cows on U.S. dairy farms-about 13 million fewer than there were in 1950. Dairy cows may be found either in herds on dairy farms where dairy farmers own, manage, care for, and collect milk from them, or on commercial farms. Herd sizes vary around the world depending on landholding culture and social structure. Dairy cow herds in the United States range in size from small farms of a dozen animals to large herds of more than 15,000. The United Kingdom dairy herd overall has nearly 2 million cows, with about 100 head reported on an average farm. In New Zealand, the average herd has more than 375 cows, while in Australia, there are approximately 220 cows in the average herd. To maintain lactation, a dairy cow must be bred and produce calves. Depending on market conditions, the cow may be bred with a "dairy bull" or a "beef bull." Female calves with dairy breeding may be kept as replacement cows for the dairy herd. If a replacement cow turns out to be a substandard producer of milk, she then goes to market and can be slaughtered for beef. Male calves can either be used later as a breeding bull or sold and used for veal or beef. Dairy farmers usually begin breeding or artificially inseminating heifers around 13 months of age. A cow's gestation period is approximately nine months. Newborn calves are removed from their mothers quickly, usually within three days, as the mother/calf bond intensifies over time and delayed separation can cause extreme stress on the calf.

Animal nutrition focuses on the dietary needs of domesticated animals, primarily those in agriculture and food production. There are seven major classes of nutrients: carbohydrates, fats, fiber, minerals, protein, vitamin, and water. These nutrient classes can be categorized as either macronutrients or micronutrients. The macronutrients are carbohydrates, fats, fiber, proteins, and water. The micronutrients are minerals and vitamins.

The macronutrients provide structural material and energy. Some of the structural material can be used to generate energy internally, and in either case it is measured in joules or calories. Carbohydrates and proteins provide 17 kJ approximately of energy per gram, while fats provide 37 kJ per gram., though the net energy from either depends on such factors as absorption and digestive effort, which vary substantially from instance to instance. Vitamins, minerals, fiber, and water do not provide energy, but are required for other reasons. A third class dietary material, fiber, seems also to be required, for both mechanical and biochemical reasons, though the exact reasons remain unclear. Molecules of carbohydrates and fats consist of carbon, hydrogen, and oxygen atoms. Carbohydrates range from simple monosaccharides to complex polysaccharides. Fats are triglycerides, made of assorted fatty acid monomers bound to glycerol backbone. Some fatty acids, but not all, are essential in the diet: they cannot be synthesized in the body. Protein molecules contain nitrogen atoms in addition to carbon, oxygen, and hydrogen. The fundamental components of protein are nitrogen-containing amino acids, some of which are essential in the sense that humans cannot make them internally. Some of the amino acids are convertible to glucose and can be used for energy production just as ordinary glucose. By breaking down existing protein, some glucose can be produced internally; the remaining amino acids are discarded, primarily as urea in urine. This occurs normally only during prolonged starvation. Other micronutrients include antioxidants and phytochemicals which are said to influence some body systems. Their necessity is not as well established as in the case of, for instance, vitamins. Most foods contain a mix of some or all of the nutrient classes, together with other substances such as toxins or various sorts. Some nutrients can be stored internally, while others are required more or less continuously. Poor health can be caused by a lack of required nutrients or, in extreme cases, too much of a required nutrient. For example, both salt and water will cause illness or even death in too large amounts. This book presents a systematic and accessible account of Dairy Chemistry and Animal Nutrition has undergone extraordinary development in the last three decades. The present book offers a compendium of reviews of the nearly all active areas of research in biochemistry, medicine and related fields of biological sciences.

I thank all members of my team who have helped in the preparation of the book. My special thanks go to "Random Publications" who have published the book.

– Shipra Chawla

Contents

1

Livestock Farming

BRIEF DESCRIPTION OF THE SECTOR

The use of cattle, sheep, goats, pigs, poultry, and other livestock offers many benefits to millions of farmers in the developing world. These animals are integral to rural livelihoods and culture, providing food (meat, blood, eggs and dairy products), materials (wool, hide, horns, etc.), income, and mechanical power for pulling carts or plowing fields. Asia has been identified as the developing region in which the demand for livestock products is expected to rise most rapidly. Livestock manure can serve as a source of fertilizer. Grazing can help sustain vegetation and promote biodiversity by dispersing seeds, controlling shrub growth, breaking soil crusts, stimulating grass growth and improving seed germination. Livestock also may represent savings and currency, and have cultural value as well. For example, gifts of livestock may serve to resolve conflicts or cement marriages.

Livestock production can be categorized under three main systems: grazing, mixed farming and industrial:

1. Grazing systems generally rely on native grassland or forests for fodder, with little or no use of crops or imported inputs, and are traditionally managed by pastoralist communities.
2. Mixed farming systems integrate livestock and crop production. Adding livestock to their farms helps farmers diversify risk and extract value from otherwise valueless or low-value by-products of each activity: crop residue becomes feed, manure becomes fertilizer. Soil nutrients can be further replenished by rotating leguminous (nitrogenfixing) fodder crops with food crops. These systems are managed by settled farmers.
3. Industrial production systems concentrate livestock populations in special facilities and separate their feeding and waste processing from the land on which they live. Feed is provided directly instead of being acquired through grazing, and manure is transported off-site. Generally, these systems are owned by relatively wealthy individuals and managed by local employees.

Grazing systems are most favoured in arid, semi-arid, or other areas of marginal value for crop-based agricultural production, while mixed farming systems flourish in temperate, subhumid, humid, and some highland climates. Industrial production, because it does not depend on local fodder supplies, can be conducted in any climate and generally occurs near the urban centers it supplies, sometimes even in peri-urban areas.

Although more slowly in sub-Saharan Africa than in most other regions. This increase is driven by growing population, increasing urbanization and rising incomes. This situation is expected to continue throughout the next decade. A shift towards industrial production—farming of monogastric species (pigs, poultry) fed with grain—may be an unavoidable trend in areas with rapidly growing demand for animal food products. Properly managed, livestock production can enhance land and water quality, biodiversity, and social and economic well-being.

However, when improperly managed, livestock production may cause significant economic, social and environmental damage. Increasing livestock production has the potential to increase environmental harm. This guideline is intended to help identify potential adverse environmental impacts and to suggest mitigation and monitoring options, as well as "best management practices" to address them.

POTENTIAL ENVIRONMENTAL IMPACTS IN THE LIVESTOCK SECTOR AND THEIR CAUSES

LARGE AREAS OF LAND DEGRADED

Overgrazing

Overgrazing of rangeland reduces the density of vegetation and the amount of organic matter generated. This, in turn, increases soil erosion from wind and water and decreases soil fertility through loss of nutrients. In arid and semi-arid areas these impacts may also contribute to desertification. Fortunately, ecosystems in these areas demonstrate considerable resilience and often recover when grazing pressure is reduced, either through traditional methods or through modern management practices.

Policy and Legal Problems

In areas that traditionally rely on grazing, the health of rangeland is generally best maintained by traditional pastoralist practices, which regulate grazing location and herd size in accordance with drought cycles and the supply of fodder. In fact, government policies or donor interventions that disrupt or discourage these practices may be a root cause of degradation. A variety of government policies may restrict the movement of livestock within a range area and prevent livestock managers from moving stock from areas that have been depleted of fodder to better supplied areas.

Two particular policy-based problems are:

1. *Land Tenure Insecurity:* Lack of confidence in secure title to rangeland (especially on communal lands) has been shown to reduce the incentive to manage the land sustainably. Many national governments have either implicitly or explicitly claimed ownership of range and wildlands and ignored traditional or customary claims.
2. *Privatization of Communal Resources:* Where national governments have privatized, or are privatizing, formerly state-owned or communal lands, new owners may erect fencing or prevent herds from crossing or grazing on their property.

Wells and Boreholes

Traditionally, access to water on critical grazing lands has been controlled to limit livestock populations and prevent herds from outgrowing the forage supply in dry areas. Thus, new wells or boreholes, generally sponsored by donors or governments, may undermine traditional livestock management systems practices by allowing herds to grow beyond sustainable levels for surrounding areas.

Overgrazing and degradation are most noticeable in the immediate vicinity of the boreholes or wells, but their effects can extend (in gradually decreasing severity) over a considerable radius. Boreholes also reduce pressure on livestock owners to decrease herd size during drought and may discourage movement of herds to other rangelands, disrupting historic wet season/dry season grazing patterns. Larger herd sizes and reductions in pastoral movement may prove to be a recipe for severe degradation of soil and vegetation.

Wet-season Grazing

Poor timing in the use of rangeland can also damage the soil. Wet-season grazing can compact the moist earth, reducing its ability to absorb moisture This increases erosion from water run-off.

Poor Balance of Livestock Species

Each species or breed of livestock has foraging preferences and will graze favoured areas and plants while neglecting others. Browsing animals, such as goats and camels, prefer the leafy tops of shrubs. By contrast, grazers tend to consume ground-level grasses and leafy plants. A poor balance between browsers and grazers can change the mix of plants in ways that degrade the area. For example, too many grazers can diminish the number and populations of herbaceous plant species and allow woody plants to become dominant, changing, possibly irreversibly, the character and utility of the ecosystem.

DAMAGED HABITAT AND REDUCED BIODIVERSITY

Livestock production can damage habitats and reduce biodiversity in wildlife and domestic stock, in vegetation, and in aquatic and wetland ecosystems.

Harm to Wildlife and Domestic Stock and Loss of Wildlife Habitat

The loss of habitat caused by livestock production in grazing and mixed farming systems may be one of the greatest threats to wildlife. Human population growth and density, and the accompanying increase in livestock, often leads producers to expand livestock grazing ranges into wild lands and the conversion of wild lands to mixed farming use.

Slaughter of Wildlife by Livestock Managers

Another danger to wildlife is intentional slaughter by livestock managers. Fear that the wildlife will prey on livestock and damage crops is a common motivation, as is the belief that the wildlife are competing with livestock for fodder, the desire to prevent spread of disease to livestock, and concern for human safety.

Extinction of Local Livestock Breeds

Systematic livestock production may result in loss of genetic diversity in livestock species. This is unfortunate because genetic diversity is a measure of a species' robustness. Local breeds may have traits conferring resistance to emergent or future pathogens, or have other favourable adaptations to local environments. The consistent replacement of local breeds with more productive imported ones can contribute to the extinction of that breed and of all the genetic diversity harbored within its population.

HARM TO VEGETATION

Clearing of Forest and Wild Lands

Vegetation is typically altered or destroyed when forests/wild lands are cleared or are burned to promote new growth. This changes local ecosystems and may contribute to global warming. Fires to burn vegetation are dangerous and degrade air quality.

Loss of Rangeland Fertility

Ironically, mixed farming systems may reduce the fertility of rangeland while helping to solve a farmland problem. Traditional farming practices cause a net loss of nutrients in farm soils; that is, when crops are harvested and sold (or even when human waste is deposited in a latrine) nutrients that make the soil fertile may be lost. Mixed farming reduces the extent of this loss, by transferring nutrients from the range to the farm in the form of manure. The gain in fertility for the farm is, of course, a net loss for rangeland. Over time, the altered nutrient balance can reduce the productive capacity of the range and/or lead to changes in the composition and density of plant species.

Loss of Farm Fertility

Damage from mixed farming often occurs when poverty and population growth pressure change the crop/grazing land ratio, where other nutrient sources are not available. As the land area available for each grazing animal shrinks, overgrazing becomes prevalent leading to soil erosion and nutrient loss. Harvesting livestock further reduces nutrients available for crop production.

Damage to Riparian Soil and Vegetation

Livestock in grazing and mixed farming systems often graze very heavily in riparian areas along streams and lakes. Results include trampling, loss of vegetation, soil disturbance, and soil compaction, erosion and/or sedimentation which can severely damage riparian habitats, increase siltation and adversely affect watersheds. Pesticide contamination from treatments to protect livestock from insectborne infections (*e.g.*, livestock "dipping") may ultimately reach the aquatic environment. Here it can be toxic to aquatic organisms, as well as people or animals who depend on these sources for drinking water.

Introduction of Invasive Plants

New breeds or fodder crops can introduce invasive non-native plants into a region. The manure, coats and hooves of newly introduced breeds can carry plant seeds. Most non-native plants are not invasive, but when they are, the results can be devastating.

Contamination from Manure

Livestock manure contains relatively high concentrations of nutrients, solids, enteric bacteria and other microorganisms, and organic material. The manure from industrial livestock operations is often discharged or "leaked" into lakes or streams, because it cannot be economically transported to replenish crop fields. When this occurs, the nutrients can cause eutrophication (rapid plant growth in water bodies), solids can create sedimentation, and organic material leads to oxygen depletion (BOD) of the water. Manure from mixed farming, if applied in a concentrated fashion, can lead to similar problems.

Degrade Water Quality and Reduce Water Supplies

Where water is scarce, either chronically or seasonally, the diversion of water to sustain livestock potentially limits its availability for other purposes. This is of particular concern in arid and semi-arid regions, where the construction of boreholes to supply livestock can lead to unsustainable withdrawal rates and the dangerous depletion of aquifer reserves.

Stockpiled manure can contaminate bodies of water, causing myriad adverse effects. These include eutrophication, oxygen depletion, sedimentation, contamination with enteric bacteria and possibly other pathogenic organisms, toxic pollution from pesticides, and contamination of groundwater and aquifers with both nitrates and pesticides. Moreover, high concentrations of nitrate in potable water supplies represent a potential health hazard, especially for children.

Harm to Human Health

Using water for farm animals may make it less available for the many uses that influence health, such as bathing, washing, cooking, and drinking. Moreover, excessive contamination by enteric microorganisms, toxic pesticides or nitrates in may render water unfit for human consumption and may be especially dangerous to children. Pesticides or other vector control treatments used on livestock represent threats to the health of livestock managers, their families, and others exposed directly or through water use. These substances may be toxic, cause birth defects, alter children's proper development, promote cancer, or slowly poison one or more organ systems.

Climate Change/Global Warming

Approximately 17 per cent of global methane is produced by livestock digestion and manure. Methane is a greenhouse gas that has 56 times more global warming potential than carbon dioxide over a 20-year time horizon. Furthermore, clearing wild or forest land for new fields reduces the land's ability to act as a carbon sink. Setting fire to vegetation to clear land or promote new growth for livestock to graze on may also contribute to global warming.

Odour

Concentrated manure stored at industrial livestock facilities can generate strong and unpleasant odours, damaging the quality of life of nearby residents. This problem is most evident when facilities are located in densely populated areas.

SECTOR PROGRAMME DESIGN—SOME SPECIFIC GUIDANCE

The following questions and suggestions are intended to help project designers and managers identify factors and practices that may cause—or prevent—adverse environmental impacts. Bear in mind that the first priority of most livestock managers and farmers is household food security and family welfare. Sustainable practices must always be balanced against these immediate demands.

CONSIDER CLIMATE, TERRAIN, AND ECOSYSTEM

Since environmental impacts from livestock production vary, depending on the specific climates, terrains and ecosystems involved, project designers need to address these characteristics during the initial design phase:

- What is the climate in the project area (arid, semi-arid, temperate, subhumid, humid)? What do historical records of rainfall patterns and flooding indicate? Is the proposed livestock management practice compatible with this climate?
- What terrains are found in the project area(s) (alluvial plain, highland, rocky desert, wetland, etc.)? Do they have any known vulnerabilities to livestock grazing? For example, are there many unprotected streams or rivers? Are there slopes with limited topsoil sensitive to erosion?
- Will the project encompass or border on protected or ecologically sensitive areas? Are there any threatened or endangered species in the area? Would the proposed project directly or indirectly threaten wildlife or native vegetation? For example, does the project require expansion of grazing into protected areas or make livestock more vulnerable to wildlife predators, triggering reprisals by farmers?

EVALUATE POLICY, LEGAL, CUSTOMARY AND CULTURAL CONTEXT

The policy, legal, and cultural contexts of a project merit attention, since, these factors may limit programme options or erect substantial barriers to success.

Policy/Legal

- Do livestock owners and managers have legal and recognized ownership and responsibility for land and grazing resources? What are the current or proposed national land tenure policies, and how are they implemented at the local level? Are they effective in encouraging sustainable management of grazing land and resources?
- What is the tenure status of current or proposed rangeland—Is it owned by individuals or the community? Does the government have claims?
- What wildlife protection laws exist and how are they implemented in the region?

Culture/Custom

- What role does livestock play in local culture and customs, and how might the proposed project affect these practices? Would the proposed project disrupt traditional grazing patterns?
- If there are customary land tenure arrangements, what are they and

how would the proposed livestock management system work within these arrangements? For example, will livestock herders—who often come into conflict with farmers, particularly during droughts—have a means of working out disputes with farmers?

- If livestock management arrangements are communal, how would these be affected by and/or affect the proposed development activities?

ASSESS CURRENT AND PROPOSED SPECIES AND BREEDS

Introduction of a new breed into an area should be approached with caution. The new breed may bring with it diseases that can decimate local livestock herds and wildlife. In addition, the foraging habits of a new breed may lead to a sharp decline in available forage and biodiversity. A new breed's reproductive habits can lead to a herd's uncontrolled growth.

Weeds can be accidentally introduced along with a new animal species, and they may displace desirable vegetation. The long term full costs and benefits of introducing a given new livestock species into a particular environment should be assessed. For example, large animals roam over extensive areas in search of food, often require a greater financial investment, can be more difficult to control, and have lower reproductive potential than small animals. It is important not to underestimate the value of breeds that are well adapted to the environment.

Livestock tend to overgraze favoured areas and plants while neglecting others. Native plants may not be able to survive heavy grazing. While unforaged plants tend to lose vigour and nutritional value as they mature. The introduction of new plant species (whether accidentally or intentionally) may quickly result in replacement of native plants. Even when grazing pressure is reduced, exotic plant species sometimes retain their dominance. Ask the following questions when introduction of a new breed or species is proposed:

About Current Species and Breeds

- Which wild and domestic species are already present in the area, and in what numbers?
- How have they been used in local farming systems and traditions?
- What are the feeding preferences of local livestock and wildlife? What is the balance between browsers and grazers? Do domestic species compete for resources with one another and with wildlife?
- Have population sizes of wild or domestic species changed recently?
- Could local breeds satisfy the project's needs?

About Proposed Species and Breeds

- If new species or breeds are being considered, how will their production complement or conflict with local species or breeds, wildlife, and other local resource users?

- How would they fit in local herding systems?
- Are they well suited to the local climate and environment?
- Are they resistant to local livestock diseases?
- Have alternative species or breeds been considered for possible introduction?

EVALUATE CURRENT AND PROPOSED LIVESTOCK MANAGEMENT PRACTICES

To maximize forage productivity, it is best to combine or alternate various livestock breeds on a range. Their differing food preferences can help to keep plants productive by minimizing overgrazing of a particular favoured area and allowing less preferred plant species time to mature. It is prudent to make superior forage available to those animals with the highest needs. When forage is limited, livestock managers may decide that young and milk-producing animals must have first access to new pastures and ranges with a wide variety of abundant forage.

Managers should investigate the value of various systems of rotating livestock. Rotation allows land to be grazed continuously throughout the year. Livestock can be rotated between fields or ranges to prevent the buildup of disease and to vary grazing pressures. Through either fencing or herding, they can be relocated into croplands to consume crop residues. Assessment of seasonal grazing patterns should include potential impact on soils.

Dry-season grazing can benefit the land by breaking up crusted soil and working seeds into the ground. By contrast, grazing on moist soil can cause considerable soil compaction, which reduces the soils ability to absorb moisture and can result in increased erosion from run-off during the rainy season. Many of the environmental impacts from livestock production are associated with particular practices of livestock managers.

Thus it is critical to understand current practices and how the proposed project might alter these practices or promote new ones:

- Who are the local community's livestock managers?
- What practices do a family or community use to control the size and composition of livestock herds?
- How do livestock managers currently control livestock movement? Will the proposed project change these movements in a way that might harm the environment?
- Does the proposed project require the construction of fences? If so, will they interfere with wildlife migration or transit of livestock belonging to other communities? Could the fences lead to overgrazing and land degradation? Will the fences be built with local materials? Would living fencing be practical? Would

solar-powered electric fencing be technically and economically feasible?

- Are streams and riverbanks currently protected from livestock damage? If the proposed project will open new areas to grazing, will water supplies need to be protected?
- Must steps be taken to prevent new livestock and associated animals (*e.g.*, dogs) from transmitting disease to wildlife? Is there a vaccination/animal disease control programme available for this purpose?
- Will the project involve construction of improvements (*e.g.*, boreholes or other infrastructure)? Could these lead to unplanned changes in herding patterns and overgrazing?

ASSESS DEMAND AND USE OF LIVESTOCK PRODUCTS

- Who is marketing livestock and livestock products?
- Is the demand for livestock products coming from local or outside populations? How rapidly is it increasing or decreasing? How stable is the demand?
- In preparing livestock products, are people using technologies which reduce impacts on the environment, open additional markets, or improve health and nutrition (of people and animals.)

ASSESS LIVESTOCK ECTOPARASITE MANAGEMENT

Epidemic and endemic diseases continue to be a major constraint to livestock productivity in large parts of the developing world. Although vaccines have controlled many of the epidemic diseases, they continue to cause severe economic losses through morbidity and mortality. These diseases include the infections caused by vector-borne haemoparasites and helminths. Existing technologies, such as chemotherapeutic agents and live vaccines that were previously successful in controlling these diseases, are no longer effective—because of acquired resistance or weakened delivery services. Appropriately designed alternatives are often lacking.

CONSIDER POPULATION PRESSURE AND DISEASE BURDEN

Two factors that may affect the outcome and impact of projects population growth and fatal or debilitating epidemic diseases. Population growth may lead to conflicting uses of grazing lands or reduce individual farms or rangeland to sizes that cannot sustain livestock. Fatal or debilitating epidemic diseases may weaken effective dissemination or replication of proper livestock management techniques.

- What is the current and projected population growth rate in the project area? How might this affect project sustainability in the future?

- What is the current extent of the HIV/AIDS epidemic in the region? How might this affect the composition of the population (size, ethnic makeup, age/gender distribution) and family structures necessary for project sustainability? How will development and livestock technical support services be affected?
- Are there other epidemic diseases in the region, such as sleeping sickness, that might adversely affect project implementation?

LIVESTOCK MANAGEMENT

LIVESTOCK PRODUCTION

Areca Sheath

Areca (*Areca catechu*) is a commercial crop and the fallen areca sheath is a good source of dry roughage to animals. The nutritional value of areca sheath is similar to paddy straw (CP, NDF, ADF) with an advantage of having less lignin, silica and more calcium, sulphur and copper. Feeding of areca sheath to sheep and dairy animals replacing paddy straw could be an alternate approach.

Lignin Degradation

Activities of lignolytic enzymes produced by immobilization on areca sheath recorded temperature and pH stabilities over a wide range, thus making it desirable for supplementation in animals.

Feed Additives

Out of the 44 plant products, *viz.* seed pulp (5), leaves (20), spices (13), citrus fruit peel (3), oils (3), and 23 plant combinations, screened for their anti-methanogenic activity using *in vitro* gas production test, a mixture of plant products and yet another combination of four plants appeared to have good potential as feed additives in the ruminants. Use of nitrate reducing bacteria as probiotic exhibited 80 to 99 per cent reduction of *in vitro* methane emission.

There was no accumulation of hydrogen, nitrate or nitrite in the incubation medium, and no adverse effect on *in vitro* true digestibility of feed. Supplementation of Chicory root powder, *Lactobacillus acidophilus* as symbiotic in the diet of dogs proved beneficial in terms of fibre utilization, calcium and lipid metabolism, hind gut fermentation and cellmediated and humoural immunity. *Helianthus tuberoses* L. *Jerusalem artichoke Hathi choke*; a novel phytogenic-origin probiotic, showed immense potential for use in the diet of dogs for enhancing their digestive health, lipid profile and immunity.

Unconventional Feed

Detoxification procedures for jatropha, *karanj* and neem seed cakes were standardized in an industrial process. In growing lambs, replacement of soybean meal with detoxified jatropha cake up to 37.5 per cent showed no adverse effect on DM intake, nutrient utilization, bloodbiochemical profile, growth rate and feed conversion efficiency. Nutritive value of *Robinia pseudoacacia* in terms of high organic matter, crude protein and low lignin content was comparable to *Grewia optiva,* and showed higher nutritional values than other tanniniferous forages. Combined treatment with curry and bel leaf powder had positive effect in restoration of cyclicity and fertility in acyclic goats and buffaloes.

Chelated Minerals

Cell-mediated immune response was significantly higher in kids given organic zinc (40 ppm zinc as zinc methionine) as compared to zinc sulphate.

Nutritional Manipulation

Feeding growing Frieswal bull calves with three different levels of rumen degradable protein and rumen undegradable protein for achieving a targeted body weight gain of 700 g and 1,000 g/day, yielded only 608 g and 772 g of average daily gain suggesting that there is a need to develop our own feeding standards for higher body weight gains.

Complete Pellets

Loin eye area increased linearly with increased level of concentrate mixture in different supplementation group. Kids fed diet without any concentrate mixture had lower yield of separated lean, fat and bone content. Different levels of supplementation improved meat bone ratio of Barbari kids under intensive system.

YAK

Two complete feed blocks (CFB) were prepared based on tree leaves (50 per cent tree leaves, 43 per cent concentrate and 7 per cent molasses) and paddy straw (50 per cent paddy straw, 43 per cent concentrate and 7 per cent molasses). The result indicated that tree leaves based CFB supplemented with limiting micronutrients can be fed during winter to overcome the weight loss in yaks.

POULTRY

Feeding Schedules for Poultry

In PB-1 females, ME restriction was successful in achieving the desired body weight (83 g/week and 5. 75 per cent growth rate) at 20 weeks of age. *Karanj* cake in the diet adversely affected Krishibro chicks, but feeding of detoxified cake improved the performance. The production performance of PB-2 line was better with organic Zn supplementation at 60 ppm and hatchability at 80 ppm. Trace mineral retention in tissues of broiler chicks was linearly related to their level in diet. Herbal antifungal compounds, *viz. Piper longum*, *Ferula assafoetida* and *Cymbopogon citratus* oil were identified as potent antifungal herbal compounds against *Aspergillus parasiticus* in poultry feed under *in vitro* conditions.

Quality Protein Maize in Diet

Dietary replacement (50 to 100 per cent w/w) of normal maize (higher in lysin and tryptophan by 44 and 33 per cent, respectively, over normal maize) resulted in 6.93 per cent higher body weight gain, 7.69 per cent improvement in feed utilization efficiency and 5.15 per cent higher breast meat yield in broiler chicken.

Amelioration of Heat Stress in Broilers

Amelioration of heat stress in broiler chickens was achieved through supplementation of vitamin C and potassium chloride in diet. Heat shock protein 70 (HSP70) was detected in muscle and blood lymphocyte in birds as potential marker of heat stress.

REPRODUCTION

Cattle

Good quality fresh and post-thaw semen of crossbred bulls had significantly higher percentage of HOST positive spermatozoa with low LDH and GOT enzyme leakage as compared to poor quality semen. Frozen semen doses (166,610) of Frieswal bulls were produced and 73,205 doses were distributed that included 50,725 to military farms, 9,000 to field progeny testing project, and 13,480 to various developmental agencies and farmers for cattle improvement.

Buffalo

Buffalo sperm membrane is highly prone for lipid peroxidation as compared to cattle, and in a study on nutrition and male fertility, it was observed that PUFA feeding may reduce lipid peroxidation and improve frozen semen fertility. Studies on nutrition-endocrine interactions in attainment of puberty in female buffaloes showed that supplementation of higher dietary energy (through bypass fat) has a positive effect on early maturity of hypothalamo-pituitary—ovarian axis in terms of early and better LH secretion and ovarian follicular activity in female buffalo calves.

Studies on physiological reasons for low reproductive efficiency and early embryonic mortality in buffaloes showed that antioxidant enzymes, Cu and Zn-SOD are induced significantly during early pregnancy and may be important for overcoming oxidative stress that otherwise would lead to early embryonic mortality. LPS and TNFα appear to modulate prostaglandin production in endometrium and corpus luteum and are important for protecting and enhancing function of corpus luteum to prevent early embryonic mortality.

IGF-I also stimulated progesterone secretion from the corpus luteum. Five proteins, specific to uterine secretions of pregnant buffalo were recognized and it may be possible to use one of these for early pregnancy diagnosis to monitor pregnancy and prevent early embryonic mortality.

Goat

Study on effect of dextran, a new cryoprotectant, revealed that dextran alone did not prove to be an efficient cryoprotectant, but 3 per cent glycerol dextran in combination with 6 per cent glycerol resulted in 39.33±1.45 per cent post-thaw recovery. Hormone delivery system for synchronization of heat in goats was standardized by using sponges of different sizes and shapes. The estrus response, onset of estrus and duration of estrus following use of intra vaginal sponge pessaries for induction of multiple birth during peak breeding activity were 100 per cent, 29.14±4.42 h and 25.71±0.37 h, respectively, with 57.14 per cent conception rate.

The kidding rate following induction of multiple birth was 2.0. During low breeding activity season (summer) the estrus response, onset of estrus and duration of estrus were 71.42 per cent, 60.0±16.54 h and 26.4±11.63 h, respectively, following use of intra vaginal pessaries. Oocytes were recovered by follicle puncture technique for *in vitro* maturation, fertilization and culture from goat ovaries. The average recovery of oocytes using follicle puncture was 1.76/ovary.

The oocytes were cultured in tissue culture medium (TCM-199) supplemented with EGF (10 ng/ml) and fertilized *in vitro* with a cleavage rate of 41.55 per cent. Embryos were bisected through microtools. Out of 25 bisected cells 3 single cells divided into double cells whereas one bunch of 3 cells developed into 4 cells. Adaptability study on goats revealed that water intake of goats kept under asbestos sheet roof sheds was significantly higher compared to the thatch roof shed during both the moderate and cool seasons. Milk production of animals kept inside the thatch roof shed was significantly higher than the animals inside the asbestos roof shed.

Camel

Physiological parameters (rectal temperature, respiration rate, pulse rate) were higher during evening as compared to morning. The morning and evening variation in the THI value was significant for all weeks in different months from October to March. The Benezara coefficient of adaptability (BCA) of camel revealed that BCA was significantly higher during evening as compared to morning. Non-pregnant female camels were divided in groups A,B,C. Each group was tested with poll gland secretion collected in breeding season (PGS-BS), poll gland secretion collected during non-breeding season (PGS-NBS) and water. The results indicated that the PGS-BS evoked greater olfactory, contact and flehmen response than PGS-NBS and water. Nine female camels, which did not conceive during breeding season were given injection of highly purified hCG for infertility treatment at the time of mating. Pregnancy was diagnosed in three out of nine camels.

Yak

On application of the vaginal sponge containing 600 mg progesterone for synchronization of oestrus followed by AI in yaks resulted in 100 per cent oestrus response with 60 per cent conception rate. The yaks do not experience cold stress (environmental stress) during winter but they are subjected to heat stress (during summer) when the ambient temperature exceeds 10°C at the altitude of 2,750 m above mean sea level. The body weight gain was significantly higher during summer than winter in adult yaks.

The lactating yak cows lost about 5.84 per cent of their body weight during winter along with reduced milk yield mainly due to shortage of feed and fodder. Estrous was displayed year round but the highest incidence (around 65 per cent) was restricted to warmer part of the year. Physiological responses, *viz.* rectal temperature, respiration and pulse rates of yak calves, bulls and lactating cows were recorded during winter and summer. The physiological responses of calves were significantly higher than bulls and lactating cows during summer. Standardization for heat shock protein 70 (Hsp 70) assay was carried out in yaks.

Mithun

Different protocols for preservation of mithun semen indicated that it is possible to preserve mithun spermatozoa at refrigeration temperature in tris-egg yolk diluents till 108 h, and doubling the concentrations of fructose in the diluent buffer increased the progressive motility, live sperm count and decreased total sperm abnormalities. Hence, addition of extra-energy source (fructose) might increase the storage time of mithun sperm at refrigerated temperature up to 5 days.

Pig

Gloved hand method of semen collection was standardized. Semen preservation was done for Hampshire, Duroc and Ghungroo pigs for first time in Asom. The AI technology was used in the farm and in the nearby villages and highest litter size (15 piglets) at birth was recorded from a Ghungroo sow in the farm.

Poultry

Exposure of birds to 650 nm and 475 nm wavelengths of light improved egg production by 6.55 and 3.05 per cent, respectively, compared to 450 nm (incandescent) light.

LIVESTOCK PROTECTION

Foot-and-mouth Disease

During the period under the report, 799 outbreaks were recorded throughout the country. Maximum outbreaks were recorded in the eastern region where there was increase in the number of outbreaks compared to last year. Disease incidence in West Bengal directly affects the situation in NE region due to animal migration. Drastic reduction in outbreaks was noticed in southern region. FMD cases were not reported in Tamil Nadu, whereas Himachal and Punjab recorded a single case of FMD each. Maximum incidence of disease was reported in March and from August to November.

Out of 1,624 clinical specimens collected from 799 outbreaks, the virus could be detected in 1,067 samples (serotype O-991, A-38 and Asia 1-38) by using sandwich ELISA and multiplex PCR. Serotype O was responsible for maximum number of outbreaks (568) followed by A (33) and Asia 1 (9). Increase in serotype O incidence was noticed in all regions except northern and western region. Serotype A, which was absent in northern and central region last year appeared this year, making it the second most prevalent serotype affecting geographically all the regions of India.

Serotype Asia 1 was completely absent in southern, central and north eastern regions. Incidence of Asia1 serotype increased in western region. Though majority of the outbreaks involved cattle, disease was also reported in buffaloes, pigs, sheep and goats. A major shift was observed in genetic lineage of serotype O viruses circulating in India. The 'Ind2001' lineage viruses gained upper hand after a gap of 8 years and outcompeted PanAsia II lineage.

The cases due to PanAsia II, PanAsia 1 and Branch C-II viruses were not completely absent during the year. Branch-C II lineage viruses responsible for outbreaks during 2008 in West Bengal were recorded in Tripura, Asom, West Bengal and Odisha during 2009–10. In serotype A, it was observed that both VP359 deletion and non deletion mutants belonging to genotype VII are cocirculating in the field in recent times.

Precisely, the viruses from Punjab and Uttarakhand shared ancestry and clustered in the same VP359 deletion lineage, VIIf, whereas viruses from two different outbreaks from Odisha and those from Andhra Pradesh clustered together and revealed no deletion in the VP3 coding region. During the period, outbreak due to Asia1 serotype was recorded in Gujarat, Madhya Pradesh, Maharashtra, Uttar Pradesh and West Bengal.

The Asia 1 field isolates were grouped with lineage C reiterating the supremacy of this lineage in the field since 2005. In vaccine matching exercise

33 virus isolates were subjected to one-way antigenic relationship study. All the isolates showed close antigenic match with respective in-use vaccine strains indicating appropriate antigenic coverage. Virus isolates (133) including 97 type O, 12 type Asia 1 and 24 type A field isolates were added.

In National FMD Virus Repository 1,687 (1,096-O, 317-Asia 1,259-A, 15-C) well characterized field isolates are available. Under National FMD sero-surveillance testing of 11,560 serum samples from 13 states revealed the presence of protective antibody titre in 42.3 per cent, 40 per cent and 30.5 per cent animal population against serotypes O, A and Asia1 respectively. This baseline response has improved due to application of vaccine. The current in-use vaccine strain IND R2/75 is found still the best and covers all the type O circulating outbreak strains in the country.

Classical Swine Fever

Classical swine fever (CSF), the topmost diseases in pigs, is one of the OIE listed diseases. The CSF or hog cholera is associated with high mortality, abortion, mummification of fetuses. It is of great concern in areas, where the pig population is in high density. CSF virus isolates maintained at the repository of PD-ADMAS, Bengaluru, were revived and processed for RT-PCR of NS5B genomic region. Phylogenetic tree based on the alignment of 408 nts of NS5B region was constructed. It showed three major branches, each representing three previously described genetic groups with further sub-branches.

The majority of the Indian CSFV isolates belong to subgroup 1.1. Clustering of the Indian field isolates away from the vaccine strain and other related strains revealed that the field isolates did not have any vaccine origin. Phylogenetic analysis based on 52 UTR classified subgroup 1.1 Indian isolates into two minor subbranches, however, no such classification could be seen from the analysis based on NS5b region; nevertheless in the latter analysis grouping of isolates was much more resolved. Lone isolate, IND294/08, is clustered within subgroup 2.2, ak into its grouping based on 52 UTR sequence.

Brucellosis

Serum samples from goat and sheep received from Gujarat, Rajasthan, Uttar Pradesh and Maharashtra revealed that among them 6.35 per cent were positive from brucellosis. Similarly, 1.8 per cent sheep and goat sera samples for Karnataka were positive for brucellosis. Porcine brucellosis is one of the most important emerging zoonoses and currently, no control programme for porcine brucellosis exists and the epidemiological situation is still unknown.

Indirect-ELISA for screening brucellosis in swine was standardized using smooth lipopolysaccahride antigen from a standard strain *Brucella abortus* S99; 22.6 per cent were found positive, which is much higher than that found in sheep and goat. An indirect ELISA was developed to detect *Brucella* in any of the susceptible species. Expression of four recombinant *Brucella* proteins, *viz.* BP26, BP39, superoxide dismutase (SOD) and *Brucella lumazine* synthase (BLS) is being envisaged to overcome the problem of cross reactivity in diagnosis of this disease.

Leptospirosis

PCR based techniques were developed for differentiation of pathogenic and nonpathogenic *Leptospira* as well as for typing. Initially, *Leptospira* genus specific PCR was carried out to differentiate the *Leptospira* from other spirochete organisms, *viz. Borrelia, Treponema* and *Leptonema* etc., which amplified 331 bp products from the *Leptospira* organism only.

Sero Prevalence of IBR

Serum samples collected from Andhra Pradesh, Gujarat, Jammu and Kashmir, Karnataka, Madhya Pradesh Maharashtra, Meghalaya, Manipur, Orissa, Punjab, Rajasthan, and Tamil Nadu, were subjected to Avidin Biotin ELISA. An overall apparent percentage positive was 34. The highest prevalence of 67 per cent was in Tamil Nadu, and the lowest in Meghalaya. Region-wise positive percentage were as follows: East zone 17 per cent, south zone 24 per cent, west zone 48 per cent, north zone 37 per cent, north-east 39 per cent, and central zone 25 per cent.

Herbal Anticoccidial and Antidiarrhoel Drugs

The extracts CIRG-5 and CIRG-6 from two different plants were evaluated *in vitro* against sporulation of coccidian oocysts and found 96 to 100 per cent effective to check the sporulation up to 96 h. Eight potential plant formulations of herbal extracts were subjected for clinical trials to control diarrhoea caused by *Escherichia coli* in kids of less than 1 month of age. Final product was developed from plant extract prototype CIRG-I and CIRG-2A and final products (formulation) @ of 10 mg/kg bwt for 1–2 days orally was highly efficacious to control EPEC diarrhoea in kid.

VACCINES AND THERAPEUTICS

- Live attenuated orf, buffalopox and camelpox vaccines met the thermo-stability and potency requirements of the biologicals.
- Bi-valent DNA vaccine containing HN and F genes of Newcastle disease virus was evaluated along with genetic adjuvant comprising IL-2, IL-4, IFNγ, and GMCSF gene constructs.
- An aroA gene knockout mutant of *Pasteurella multocida* serotype B:2 was constructed for the development of HS mutant vaccine.
- Chitosan coated cationic PLG nanoparticles were prepared and when tested for the transfection efficiency with commercial transfecting agent, indicated that these nanoparticles can be safely used for delivery of DNA vaccines.

- *Trypanosoma evansi* beta tubulin protein was expressed in both prokaryotic and eukaryotic expression systems.
- Two new target proteins of *Fasciola gigantica, viz.* leucine aminopeptidase and peroxiredoxin coding genes were cloned and successfully expressed in their native conformation in a prokaryotic expression system.
- Immunomodulators of microbial and herbal origin, *i.e. Mycobacterium phlei, Lactobacillus acidophilus* and *Tinospora cordifolia* were effective under field conditions to improve productive performance and mitigate the adverse effect of sub-clinical caecal coccidiosis and thereby improvement of body weight gain in broiler birds.
- Pharmacokinetics of gatifloxacin revealed that it has therapeutic potential for use in sheep.
- Medetomidine-butorphanol, thiopental and halothane combination provided better clinicophysiological and haemodynamic stability in general anaesthesia in buffaloes.
- Three TLR genes of yak (TLR 5, 7 and 9) and 2 TLR genes of mithun (TLR 4 and 7) were sequenced completely.

DIAGNOSTICS

- A recombinant lipoprotein 'P48' of *Mycoplasma agalactiae* was evaluated in an ELISA as a rapid immunodiagnostic tool in sheep and goats.
- A genus-specific PCR assay was standardized for rapid detection of *Arcobacter* sp. targeting 16S rRNA for confirmation of the isolates of this emerging pathogen.
- A 36 kDa protein was identified in different serovars of *Salmonella* during biofilm formation.
- Magnetic nanoparticles (MNPs) were synthesized by two different processes and were characterized for structural, morphological, size and magnetic properties. The MNPs were functionalized for immunoassay applications and evaluated for cellular toxicity in *in-vitro* experiments.
- A specific and sensitive diagnostic marker was identified for early detection of pancreatitis in dog.

MOLECULAR CHARACTERIZATION OF PATHOGENS/RECEPTORS

- Indian H5N1 isolates were characterized genetically and grouped with EMA *3/2.2.3 subclade,* which branched from two distinct nodes in the phylogenetic tree.
- The alignment of deduced amino acid sequences showed that all the Indian H5N1 viruses characterized contained the multiple basic amino acid motif at their HA cleavage site, which is characteristic of highly pathogenic avian influenza viruses.

- Two viruses from 2010 outbreaks in West Bengal contained amino acid asparagine at position 31 of the M2 protein, suggesting that both viruses are resistant to drug amantadine.
- An isolate of buffalo pox from man and animal recovered from an outbreak from Pune was characterized for B5R, H3L, D8L and A27L genes.
- An orf isolate from camel in a sporadic infection was characterized for B2L, GIF and VLTF-4 genes.
- Genotyping and partial sequencing of rotaviruses from man and pigs revealed human-pig reassortants as well as presence of novel combinations in India.
- Complete HN gene of velogenic Newcastle disease virus containing kozak sequence was amplified and cloned into pcDNA3. 1 eukaryotic expression for enhanced expression.
- Phylogenetic analysis by AP-PCR, PFGE and ribotyping showed considerable similarity between human and fish isolates of pathogenic and pandemic *Vibrio parahaemolyticus,* which reflected the public health risk potentials of fish.
- Molecular characterization by PCR-SSCP analysis of *Oesophagostomum* spp. revealed two distinct conformers with six nucleotide polymorphisms (SNPs) among male and female parasites.
- Probability of evolutionary changes among prevailing genotypes of *Oesophagostomum* spp. was predicted by analysing secondary structure of RNA.

QUALITY CONTROL AND PRODUCTION OF VETERINARY BIOLOGICALS

- 223,500 doses of RD 'F' strain vaccine; 166,000 RD 'Mukteswar' strain vaccine; 132,500 doses of fowl pox vaccine; 184,195 doses of lapinized swine fever vaccine; 423,800 doses of tissue culture sheep pox vaccine; 160,600 doses of PPR vaccine; 57,925 doses of *Brucella abortus* strain-19 (live) vaccine; 3,640 ml HS adjuvant vaccine; 73,500 doses of tuberculin PPD; 50,000 doses of Johnin PPD; 25,250 doses of mallein PPD; 55,250 ml of *Brucella* agglutination test antigen; 6,900 ml of *Brucella abortus* Bang Ring antigen; 19,860 ml of Rose Bengal Plate Test antigen; 219 ml of *Brucella* positive serum; 9,000 ml of *S.* Pullorum coloured antigen; 45 ml *S.* Pullorum positive serum; 87 ml *Salmonella* 'O' serum and 3,750 ml of *S.* Abortus equi 'H' antigen were produced and quality tested.
- 7.32 million monovalent doses of FMD vaccine were produced at Bengaluru campus.
- 435,100 doses of PPR vaccine were produced; similarly, 33 PPR c-ELISA, 22 PPR s-ELISA and 100 doses of goat pox vaccine were supplied by Mukteswar campus.

Sero-monitoring of Important Equine Diseases

During 2009–10, equine disease sero-survey was conducted in Rajasthan, Haryana, Punjab, Uttar Pradesh, Uttarakhand, Madhya Pradesh, Chhattisgarh and Sikkim. A total of 176 out of 1,045 equines (16.84 per cent) had antibodies to equine influenza virus while 28 (2.67 per cent) were positive for EHV-1 and 107 (10.43 per cent) were positive for antibodies to Japanese encephalitis virus. Antibodies to *Babesia equi* were detected in 320 (38.09 per cent) of 840 equines, while *T. evansi* antibodies were detected in 49 (5.4 per cent) out of 906 equines tested. One case of equine infectious anemia was detected in Uttarakhand. EIA-positive equine was reported after a gap of 11 years from India. None of the 1,045 equines tested showed antibodies to *Brucella* and *Salmonella* Abortus equi 'H' antigen.

Japanese Encephalitis (JE) Virus Isolation

JE, a mosquito-transmitted disease caused by JE virus, is causing disease in equines, pigs and humans. It is prevalent in eastern and southern Asia. Serum samples from clinically affected horses and pooled mosquito vector samples from the surrounding field were found positive for JEV RNA by RT-PCR. Virus was isolated from one of the horses. This is the first report of isolation of JEV from a horse in India. The phylogenetic analysis of the envelope gene (1500 bp) sequence of the isolated virus, JE/eq/India/H225/2009 indicated that the isolate belonged to genogroup III of JEV, and that the equine isolate clustered together with Vellore group of JE isolates from India.

Camelpox Virus Zoonosis

Camelpox virus (CMLV) causes skin eruptions in addition to severe generalized exanthema mostly in camels and outbreaks occur frequently in camel-rearing north-western region of the country. However, their zoonotic potential has not been proven conclusively. The disease was confirmed on the basis of isolation of the virus from diseased camels, detection of the CMLV antibodies in the serum of diseased camels and human patients by serum neutralization test and amplification of CMLV-specific three full-length envelope protein genes (A27L, H3L and D8L) and one partial gene (C18L) of the isolated viruses from camels and human specimens.

The phylogenetic analysis indicated that CMLV isolates cluster closely with variola virus (VARV) and vaccinia virus (VACV). The CMLV zoonosis indicates the changing pattern of host specificity of poxviruses and certainly points towards a declining cohort immunity against Orthopoxviruses in humans, which could be a public health concern. This is the first confirmed report of camelpox virus zoonosis.

Cysteine Proteinases from Trypanosoma Evansi

Parasite derived cysteine proteinases were proposed as potential targets for immunotherapeutic, chemotherapeutic and diagnostic reagents for parasitic diseases. The proteinases of camel isolate of *T. evansi* were analysed by gelatin substrate SDS-PAGE containing collagen antigen confirming the presence of cysteine proteinases in *T. evansi*. Some of these cysteine proteinases were reactive with experimental/field serum sample from equines with *T. evansi* infection. This indicated the potential of these enzymes in serodiagnosis of this economically important disease of livestock.

FISH

Capture Fisheries

The marine fish landings of India during the year 2009 has, provisionally, been estimated as 3. 16 million tonnes. The pelagic finfishes constituted 52 per cent, demersal fishes 28 per cent, crustaceans 16 per cent and molluscs 4 per cent of the total landings. The sectorwise contributions during the year 2009 were mechanized 74 per cent, motorized 22 per cent and artisanal 4 per cent. The west coast accounted for 56 per cent of the total landings. The northeast region, comprising West Bengal and Odisha contributed 20 per cent to the total production; southeast region consisting of Andhra Pradesh, Tamil Nadu and Puducherry contributed 24 per cent. On the west coast, the northwest region comprising Maharashtra and Gujarat recorded 26 per cent of the total, and the southwest region comprising Kerala, Karnataka and Goa contributed 30 per cent.

Cruzing for Hilsa Fisheries in Hooghly

The investigations on distribution of hilsa in Hooghly from Nabadweep to Kakdweep estuary were done through cruzing. It established that depth and width of the river stretch affect the fishing effort and hilsa catch. The zones with <3 m depth had rare fishing activity, while fishing intensity was higher in deeper water (10 m depth). Hilsa was evenly distributed over the observed >100 km freshwater stretch. Movement of hilsa in shoals was missing during non-spawning season. Hilsa was available throughout the year with peak in monsoon. The observations may lead to possible occurrence of a resident stock of hilsa in Hooghly in addition to regular seasonal influx of migrant groups from sea to freshwater zone for breeding.

Impact Assessment of Fishing Ban

To assess the impact of fishing ban on reservoir ecosystem, massbalanced models were used on different ecological groups before and after introduction of fishing ban in Wyra reservoir of Andhra Pradesh using Ecopath software. It is the first attempt in India on modeling a tropical productive reservoir, which contributes to fishery, particularly the commercially important prawn fishery.

The results of the assessment indicated:

- Slight increase in system turnover,
- Better organization/increased efficiency,
- Decreased tendency for eutrophication (low sum of all flows into detritus),
- Superior stability of total biomass/TST,
- Higher total primary production/total respiration,
- Increased resilience (higher overhead per cent), and
- Higher maturity and stability (higher primary production/biomass and lower ascendancy per cent).

Formulation of Shrimp Feed

Yard trials revealed that 3 per cent Ca with 2 per cent P supplementation resulted in a significantly higher digestibility of Ca and P as well as weight gain compared to other groups receiving lower levels of Ca and P. Feeds for high salinity, indicated that high growth can be achieved at 40 and 33.5 per cent dietary crude protein levels compared to lower dietary protein levels. Based on the better protein efficiency ratio and apparent protein utilization at high salinity (40 per cent), protein utilization efficiency was better at 33.5 per cent dietary protein level.

Reducing Fish Meal in Shrimp Feed

One of the bottlenecks to the approach of using plant proteins for reducing the level of costly fish meal in shrimp diet, is the high cellulose content of these feeds. *Bacillus macerans,* a potential cellulolytic bacteria, isolated from gut of *Lates calcarifer,* was identified for testing as feed supplement in plant protein based shrimp feed.

After 48 h of incubation, crude protein content, free glucose concentration and microbial count were maximum in the feed and dry matter loss was less. Growth, survival, protein efficiency ratio were significantly higher and feed conversion ratio was significantly lower in shrimps fed with feed supplemented with live cellulolytic bacteria as compared to control feed.

FISH FEED FOR MARINE ORNAMENTAL FISHES

The CMFRI has launched a dry formulated feeds named CadalminTM *Varna* for marine ornamental fish like clown and damsel fish. These feeds contain 40 per cent protein, 9 per cent fat, 39 per cent carbohydrates, 7 per cent ash and less than 2 per cent fibre, colour imparting nutrients like carotenoids from natural sources, nutraceuticals and microbial products. They are slow sinking crumbles available in three particle sizes, 0.25 mm, 0.75 mm and 1 mm produced through twin-screw extrusion technology, which is the state-of-the-art in aquatic feed production. The feed costs only ₹400/kg in comparison to imported feed having a price in the range of ₹4,000/kg.

Adoption of Biosecurity and WSSV Incidence

Disease surveillance in the two types of culture systems in West Bengal, shrimp monoculture and traditional bheries over a period of three years (2007–10) indicated that white spot disease (WSD), shell associated problems and gill associated problems were statistically significant among the two culture systems. Newly emerging diseases such as monodon slow growth syndrome was observed in both culture systems, whereas, white faecal disease was observed only in shrimp monoculture system.

WSD outbreak was significantly high where stocking was done by the seed, which has not been tested by PCR. Analysis of data regarding adoption rate of biosecurity protocols either single or in combination in shrimp farms in Chidambaram and Nagapattinam Districts of Tamil Nadu indicated that white spot disease incidence is at a higher rate when protocols of disinfection of implements and reservoir ponds, were not followed.

Diagnostics for Monodon Slow Growth Syndrome

One of the emerging shrimp diseases is the monodon slow growth syndrome (MSGS) with abnormal slow growth and marked size differences among individuals. An improved diagnostic nested RT-PCR with custom designed primer targeting RdRp gene of Laem-Singh virus (LSNV), which has been implicated in monodon slow growth syndrome (MSGS) was developed. Screening of farmed and wild broodstock samples with this improved diagnostic PCR showed a high prevalence of LSNV. High prevalence in broodstock could lead to the spread of this virus by vertical transmission.

2

Advanced Dairy and Food Packaging

INTRODUCTION

Food and beverage packaging comprises 55 per cent to 65 per cent of the $130 billion value of packaging in the United States. Food processing and packaging industries spend an estimated 15 per cent of the total variable costs on packaging materials. Industrial processing of food, reduced consumption of animal protein, importation of raw materials and ingredients to be converted in the United States, and scarcity of time to select/prepare food from fresh ingredients have enhanced innovation in food and beverage packaging.

The continued quest for innovation in food and beverage packaging is mostly driven by consumer needs and demands influenced by changing global trends, such as increased life expectancy, fewer organizations investing in food production and distribution, and regionally abundant and diverse food supply. The use of food packaging is a socioeconomic indicator of increased spending ability of the population or the gross domestic product as well as regional food availability.

This Scientific Status Summary provides an overview of the latest innovations in food packaging. It begins with a brief history of food and beverage packaging, covering the more prominent packaging developments from the past, and proceeds to more modern advances in the packaging industry.

The article then delves into current and emerging innovations in active and intelligent packaging (such as oxygen scavengers and moisture control agents), packaging mechanisms that control volatile flavours and aromas (such as flavour and odour absorbers), and cutting-edge advances in food packaging distribution (such as radio frequency identification and electronic product codes). Finally, the nano-sized components that have the potential to transform the food packaging industry.

HISTORY OF FOOD AND BEVERAGE PACKAGING

Modern food packaging is believed to have begun in the 19th century with the invention of canning by Nicholas Appert. After the inauguration of food microbiology by Louis Pasteur and colleagues in the 19th century, Samuel C. Prescott and William L. Underwood worked to establish the fundamental principles of bacteriology as applied to canning processes. These endeavors to preserve and package food were paralleled by several other packaging-related inventions such as cutting dies for paperboard cartons by Robert Gair and mechanical production of glass bottles by Michael Owens.

In the beginning of the 20th century, 3-piece tin-plated steel cans, glass bottles, and wooden crates were used for food and beverage distribution. Some food packaging innovations stemmed from unexpected sources. For example, Jacques E. Brandenberger's failed attempts at transparent tablecloths resulted in the invention of cellophane. In addition, wax and related petroleum-based materials used to protect ammunition during World War II became packaging materials for dry cereals and biscuits.

Many packaging innovations occurred during the period between World War I and World War II; these include aluminum foil, electrically powered packaging machinery, plastics such as polyethylene and polyvinylidene chloride, aseptic packaging, metal beer cans, flexographic printing, and flexible packaging. Most of these developments helped immeasurably in World War II by protecting military goods and foods from extreme conditions in war zones.

Tin-plated soldered side-seam steel progressed to welded sideseam tin-free steel for cans, and 2-piece aluminum with easy open pop tops were invented for beverage cans, spearheading the exponential growth of canned carbonated beverages and beer during the 1960s and 1970s. The development of polypropylene, polyester, and ethylene vinyl alcohol polymers led the incredible move away from metal, glass, and paperboard packaging to plastic and flexible packaging.

Later 20th century innovations include active packaging (oxygen controllers, antimicrobials, respiration mediators, and odour/aroma controllers) and intelligent or smart packaging. Distribution packaging is already influenced by the potential role of radio frequency identification for tracking purposes. Products such as retort pouches and trays, stand-up flexible pouches, zipper closures on flexible pouches, coextrusion for films and bottles, and an inexorable drive by injection stretch blowmolded polyester bottles and jars for carbonated beverages and water have emerged as rigid and semi-rigid packaging.

Multilayer barrier plastic cans, microwave susceptors, dispensing closures, gas barrier bags for prime cuts of meat, modified atmosphere packaging, rotogravure printed full-panel shrink film labels, and dual oven able trays are examples of innovations for the convenience attributes that have propelled food and beverage packaging into the 21st century. Moreover, some 21st century innovations are related to nanotechnology whose future may lie in improving barrier and structural/mechanical properties of packaging materials and development of sensing technologies. The principal drivers for most of these innovations have been consumer and food service needs and demands for global and fast transport of food. These packaging innovations are derived largely from industry research and development programmes.

THE EXPANDED ROLES OF FOOD AND BEVERAGE PACKAGING

The principal function of packaging is protection and preservation from external contamination. This function involves retardation of deterioration, extension of shelf life, and maintenance of quality and safety of packaged food. Packaging protects food from environmental influences such as heat, light, the presence or absence of moisture, oxygen, pressure, enzymes, spurious odours, microorganisms, insects, dirt and dust particles, gaseous emissions, and so on. All of these cause deterioration of foods and beverages.

Prolonging shelf life involves retardation of enzymatic, microbial, and biochemical reactions through various strategies such as temperature control; moisture control; addition of chemicals such as salt, sugar, carbon dioxide, or natural acids; removal of oxygen; or a combination of these with effective packaging. Precise integration of the product, process, package, and distribution is critical to avoid recontamination. The ideal packaging material should be inert and resistant to hazards and should not allow molecular transfer from or to packaging materials.

Other major functions of packaging include containment, convenience, marketing, and communication. Containment involves ensuring that a product is not intentionally spilled or dispersed. The communication function serves as the link between consumer and food processor. It contains mandatory information such as weight, source, ingredients, and now, nutritional value and cautions for use required by law. Product promotion or marketing by companies is achieved through the packages at the point of purchase.

Secondary functions of increasing importance include traceability, tamper indication, and portion control. New tracking systems enable tracking of packages though the food supply chain from source to disposal. Packages are imprinted with a universal product code to facilitate checkout and distribution control. More recent innovations used include surface variations sensed by finger tips and palms, sound/music or verbal messages, and aromas emitted as part of an active packaging spectrum. Gloss, matte, holograms, diffraction patterns, and flashing lights are also used.

ACTIVE AND INTELLIGENT FOOD PACKAGING

Traditional food packages are passive barriers designed to delay the adverse effects of the environment on the food product. Active packaging, however, allows packages to interact with food and the environment and play a dynamic role in food preservation. Developments in active packaging have led to advances in many areas, including delayed oxidation and controlled respiration rate, microbial growth, and moisture migration. Other active packaging technologies include carbon dioxide absorbers/emitters, odour absorbers, ethylene removers, and aroma emitters. While purge and moisture control and oxygen removal have been prominent in active packaging, purge control is the most successful commercially.

An example is the drip-absorbing pad used in the poultry industry. In addition, active packaging technology can manipulate permselectivity, which is the selective permeation of package materials to various gases. Through coating, microperforation, lamination, coextrusion, or polymer blending, permselectivity can be manipulated to modify the atmospheric concentration of gaseous compounds inside a package, relative to the oxidation or respiration kinetics of foods.

Certain nanocomposite materials can also serve as active packaging by actively preventing oxygen, carbon dioxide, and moisture from reaching food. Intelligent or smart packaging is designed to monitor and communicate information about food quality. Examples include time-temperature indicators (TTIs), ripeness indicators, biosensors, and radio frequency identification. These smart devices may be incorporated in package materials or attached to the inside or outside of a package.

As of summer 2008, the commercial application of these technologies to food packaging has been small. However, the U.S. Food and Drug Administration (FDA) recognizes TTIs in the 3rd edition of the *Fish and Fisheries Products Hazards and Control Guidance*, so their importance may increase in the seafood industry. Moreover, Wal-Mart, Home Depot and other retail outlets use radio frequency identification, so it is likely to become very prominent as a mechanism for tracking and tracing produce and others perishable commodities.

OXYGEN SCAVENGERS

The presence of oxygen in a package can trigger or accelerate oxidative reactions that result in food deterioration: Oxygen facilitates the growth of aerobic microbes and molds. Oxidative reactions result in adverse qualities such as off-odours, off-flavours, undesirable colour changes, and reduced nutritional quality. Oxygen scavengers remove oxygen (residual and/or entering), thereby retarding oxidative reactions, and they come in various forms: sachets in headspace, labels, or direct incorporation into package material and/or closures.

Oxygen scavenging compounds are mostly agents that react with oxygen to reduce its concentration. Ferrous oxide is the most commonly used scavenger. Others include ascorbic acid, sulfites, catechol, some nylons, photosensitive dyes, unsaturated hydrocarbons, ligands, and enzymes such as glucose oxidase. To prevent scavengers from acting prematurely, specialized mechanisms can trigger the scavenging reaction. For example, photosensitive dyes irradiated with ultraviolet light activate oxygen removal. Oxygen scavenging technologies have been successfully used in the meat industry.

CARBON DIOXIDE ABSORBERS AND EMITTERS

Carbon dioxide may be added for beneficial effects, for example, to suppress microbial growth in certain products such as fresh meat, poultry, cheese, and baked goods. Carbon dioxide is also used to reduce the respiration rate of fresh produce and to overcome package collapse or partial vacuum caused by oxygen scavengers. Carbon dioxide is available in various forms, such as moisture-activated bicarbonate chemicals in sachets and absorbent pads. Conversely, high levels of carbon dioxide resulting from food deterioration or oxidative reactions could cause adverse quality effects in food products. Excess carbon dioxide can be removed by using highly permeable plastics whose permeability increases with higher temperatures.

MOISTURE CONTROL AGENTS

For moisture-sensitive foods, excess moisture in packages can have detrimental results: for example, caking in powdered products, softening of crispy products such as crackers, and moistening of hygroscopic products such as sweets and candy. Conversely, too much moisture loss from food may result in product desiccation.

Moisture control agents help control water activity, thus reducing microbial growth; remove melting water from frozen products and blood or fluids from meat products; prevent condensation from fresh produce; and keep the rate of lipid oxidation in check. Desiccants such as silica gels, natural clays and calcium oxide are used with dry foods while internal humidity controllers are used for high moisture foods (for example, meat, poultry, fruits, and vegetables).

Desiccants usually take the form of internal porous sachets or perforated water-vapour barrier plastic cartridges containing desiccants. They can also be incorporated in packaging material. Humidity controllers help maintain optimum in-package relative humidity (about 85 per cent for cut fruits and vegetables), reduce moisture loss, and retard excess moisture in headspace and interstices where microorganisms can grow. Purge absorbers remove liquid squeezed or leaking from fresh products and can be enhanced by other active additives such as oxygen scavengers, antimicrobials, pH reducers, and carbon dioxide generators.

ANTIMICROBIALS

Antimicrobials in food packaging are used to enhance quality and safety by reducing surface contamination of processed food; they are not a substitute for good sanitation practices. Antimicrobials reduce the growth rate and maximum population of microorganisms (spoilage and pathogenic) by extending the lag phase of microbes or inactivating them. Antimicrobial agents may be incorporated directly into packaging materials for slow release to the food surface or may be used in vapour form.

Research is underway on the antimicrobial properties of the following agents:

- *Silver Ions*: Silver salts function on direct contact, but they migrate slowly and react preferentially with organics. Research on the use of silver nanoparticles as antimicrobials in food packaging is ongoing, but at least 1 product has already emerged: FresherLongerTM storage containers allegedly contain silver nanoparticles infused into polypropylene base material for inhibition of growth of microorganisms.
- *Ethyl Alcohol*: Ethyl alcohol adsorbed on silica or zeolite is emitted by evaporation and is somewhat effective but leaves a secondary odour.
- *Chlorine Dioxide*: Chlorine dioxide is a gas that permeates through the packaged product. It is broadly effective against microorganisms but has adverse secondary effects such as darkening meat colour and bleaching green vegetables.
- *Nisin*: Nisin has been found to be most effective against lactic acid and Gram-positive bacteria. It acts by incorporating itself in the cytoplasmic membrane of target cells and works best in acidic conditions.
- *Organic Acids*: Organic acids such as acetic, benzoic, lactic, tartaric, and propionic are used as preservative agents.
- *Allyl Isothiocyanate*: Allyl isothiocyanate, an active component in wasabi, mustard, and horseradish, is an effective broad spectrum antimicrobial and antimycotic. However, it has strong adverse secondary odour effects in food.
- *Spice-based Essential Oils*: Spice-based essential oils have been studied for antimicrobial effects: for example, oregano oil in meat, mustard oil in bread, oregano, basil, clove, carvacol, thymol, and cinnamon.
- *Metal Oxides*: Nanoscale levels of metal oxides such as magnesium oxide and zinc oxide are being explored as antimicrobial materials for use in food packaging.

ETHYLENE ABSORBERS AND ADSORBERS

Ethylene is a natural plant hormone produced by ripening produce. It accelerates produce respiration, resulting in maturity and senescence. Removing ethylene from a package environment helps extend the shelf life of fresh produce. The most common agent of ethylene removal is potassium permanganate, which oxidizes ethylene to acetate and ethanol. Ethylene may also be removed by physical adsorption on active surfaces such as activated carbon or zeolite. Potassium permanganate is mostly supplied in sachets while other adsorbent or absorbent chemicals may be distributed as sachets or incorporated in the packaging materials.

TEMPERATURE CONTROL: SELF-HEATING AND COOLING

Self-heating packaging employs calcium or magnesium oxide and water to generate an exothermic reaction. It has been used for plastic coffee cans, military rations, and on-the-go meal platters. The heating device occupies a significant amount of volume (almost half) within the package. Self-cooling packaging involves the evaporation of an external compound that removes heat from contents (usually water that is evaporated and adsorbed on surfaces).

ADVANCES IN CONTROLLING VOLATILE FLAVOURS AND AROMAS

The mass transfer of components between and within food and packaging leads to the loss of volatile flavours and aromas from food. The most common methods of mass transfer food packaging systems are migration, flavour scalping, selective permeation, and ingredient transfer between heterogeneous parts of the food.

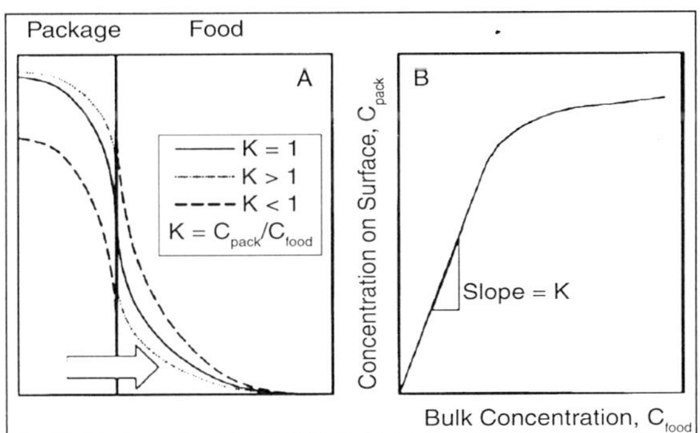

Fig. Effect of Interfacial Partitioning of Molecules on Mass Transfer between Food and Packaging Materials

Notes:

(a) Migration of Package Constituents into Foods,

(b) Scalping of Food Flavour into Packaging Materials. K is a Partition Coefficient of a Migrant

Migration is the transfer of substances from the package into the food due to direct contact. Migration of packaging components to food must be understood and considered with toxicological risk analysis. Most incidences of migration occur in plastic packaging systems; thus, the most commonly studied migrants are plastic monomers, dimers, oligomers, antioxidants, plasticizers, and dye/adhesive solvent residues. Migration of packaging material components is examined in 2 ways, based on the migrating chemicals.

One is global migration (that is, total migration); the other is specific migration of chemicals of interest. Migration of chemical substances is determined by the units of mg/kg for food or mg/m^2 for package surface. The degree of migration depends on several variables: contact area between food and package material, contact time, food composition, concentration of migrant, storage temperature, polymer morphology, and polarity of polymeric packaging materials and migrants. Flavour scalping is caused by the absorption of desirable volatile food flavours by package materials (for example, absorption of volatile flavours of orange juice and citrus beverages by polyethylene).

Polyethylene materials are known to scalp many volatiles from food. This is due to polyethylene's lipophilic nature, which attracts large amounts of nonpolar compounds such as volatile flavours and aroma in foods. In fact, certain products—especially high-fat foods or vacuum-packaged foods—pick up odours from adjacent strong odour foods when stored or distributed in the same case, storage room, or trailer. The absorption of undesirable flavours by packaging materials follows the same theory and principles of migration but is generally not considered flavour scalping. Unacceptable odour pick-up can be avoided by proper package wrapping with high-barrier materials.

The use of high-barrier packaging materials can also prevent the absorption of other nonfood odours such as taints. Because they result in deterioration of quality and consumer preference of the packaged food, both migration and flavour scalping are unfavourable. However, some applications intentionally utilize these methods of mass transfer to improve the quality of packaged foods. The interactions can be used in active and intelligent packaging applications.

Examples are off-flavour absorbing systems and beneficial volatile release systems. Besides loss of flavour in food, nonpolar flavour components loosen the polymer structure to create more amorphous polymers. This causes undesired changes in the mechanical (seal strength, loss of laminations) and barrier (to oxygen, moisture, and volatiles) properties. Therefore, the effect of off-flavour absorption on essential characteristics of plastic packaging materials should be investigated.

FLAVOUR AND ODOUR ABSORBERS

In packaged foods, flavour and odour absorbers take in unwanted gaseous molecules such as volatile package ingredients, chemical metabolites of foods, microbial metabolites, respiration products, or off-flavours in raw foods. Examples are sulfurous compounds and amines produced biochemically from protein degradation, aldehydes and ketones produced from lipid oxidation or anaerobic glycolysis, and bitter taste compounds in grapefruit juice.

Flavour-and odour-absorbing systems use the same mass transfer mechanism as flavour scalping to remove off-characteristics and are typically available as films, sachets, tapes/labels, and trays.

Flavour and odour absorbers are usually placed inside packages or combined with other flavour permeable materials. Although scavenging of malodourous constituents is recommended to improve the quality of packaged foods, this technology should not be used to mask off-odours produced by hazardous microorganisms that could place consumers at risk. Absorption is the dominant mass transfer phenomenon when porous materials absorb flavour molecules.

Porous or fineparticulate materials have extremely large surface areas; therefore, even with low absorption rates, the overall absorption is very significant. Porous materials—such as zeolites, clays, molecular sieves, active carbon, maltodextrin, and cyclodextrin—have been found to increase flavour. Besides the absorption rate, the maximum absorption, which is also a significant factor, can be controlled by the thickness of films, with thicker films absorbing more flavours. The film thickness does not alter absorption kinetics but relates to the total amount of absorption.

The overall absorption rate that includes the surface adsorption rate and in-structure diffusion rate is affected by molecular size of flavours, polarity of flavour molecules and packaging materials, porosity of packaging materials, storage temperature, morphology of plastic packaging materials (that is, glass transition temperature, crystallinity, and free volume), and relative humidity.

Polar plastic materials or corona treatment of film surface and/or high-barrier (or high-density and more-crystalline) materials can be used to prevent nonpolar flavour absorption. When nonporous plastic materials absorb flavour molecules without the occurrence of chemical reactions, the increase in adsorption kinetics and diffusion kinetics will accelerate the absorption rate.

However, when the flavour absorption results in chemical reactions with the plastic material, reaction kinetics as well as adsorption and diffusion kinetics significantly affect the off-flavour elimination capacity. Alkali chemicals such as ammonia and amines can be absorbed and neutralized by packaging materials containing acidic ingredients. This acid–alkali reaction mechanism between volatile chemicals and polymer ingredients has been utilized to eliminate various alkali chemicals and aldehydes.

HIGH CHEMICAL BARRIER MATERIAL INNOVATIONS

High-barrier packaging can significantly reduce adsorption, desorption, and diffusion of gases and liquids to maintain the quality of food. It also prevents the penetration of other molecules such as oxygen, pressurized liquid or gas, and water vapour, which are generally undesirable for food preservation. There are various procedures to enhance the barrier property of packaging materials or packages.

Barrier properties can be improved by combining the package materials with other high-barrier materials through polymer blending, coating, lamination, or metallization. The morphology of the blend relates to its permeability. Laminar structure (such as coating or lamination) of high-barrier materials on packaging material decreases the permeability linearly with respect to the square thickness; also, blending with platelets or droplets of high-barrier materials reduces permeability but is less effective than coating or lamination at the same mass as that of highbarrier materials. Commercial examples are polyvinylidene chloride (PVdC) coating on oriented polypropylene (OPP), polyethylene terephthalate (PET) lamination on coextruded polypropylene/polyethylene, and aluminum-metallization on PET.

Table. Oxygen Transmission Rate (OTR in $cm^3\ m^{-2}\ d^{-1}\ atm^{-1}$ at 23 °C, 50% RH) and Water Vapour Transmission Rate (WVTR in $g\ m^{-2}\ d^{-1}$ at 23 °C, 75% RH) of Composite Films based on 12 μm PET Films

Film Specification (μm)	OTR	WVTR	
PET	110	15	12
PET/PE	0.93 to 1.24	0.248 to 0.372	12/50
PET/PVDC/PE	0.33	0.132	12/4/50
PET/PVAL/PE	0.13	0.26 to 0.39	12/3/50
PET/EVOH/PE	0.06	0.134 to 0.268	12/5/50
PET/Al-met/PE	0.06 to 0.12	0.006 to 0.03	12/-/50
PET/SiO_x	0.006 to 0.06	0.0024 to 0.06	12/-
PET/Al-foil/PE	0	0	12/9/50

Notes:

PE = polyethylene low density; PVDC = poly(vinylidene chloride); PVAL =

poly(vinyl alcohol); EVOH = ethylene vinyl alcohol; Al-met = aluminum metallization; SiO_x = silicon oxide; Al-foil = aluminum foil.

Table shows barrier properties of commercially available laminated (or coated) PET films. Orientation of crystalline polymers such as bi-axially oriented polypropylene (BOPP) and biaxially oriented PET (BOPET) produces lamella structure of polymers. Other examples are blends of polymers with planar clay particles, a lamella blend structure, and a mixture of beeswax in edible polymer as particulate system films. Other innovative technologies for barrier property enhancement with commercial feasibility include transparent vacuum-deposited or plasma-deposited coating of silica oxide on PET films, epoxy spray on PET bottles, and composites of plastics with nanoparticles.

FACTORS AFFECTING FOOD PACKAGING INTERACTIONS AND BARRIER PROPERTIES

Regardless of the direction of mass transfer (for example, whether flavour absorption or release) or the intent of the mass transfer (for example, whether to achieve desirable transfer or to prevent an undesirable transfer), various factors of food, packaging, and distribution affect the mass transfer kinetics and amount. The transfer of chemicals is principally derived by the concentration difference in food and packaging materials. A small concentration gradient results in a little transfer while a large gradient results in transfer of a large amount of the compound at a fast rate.

The nature of food is also an important factor: for example, food ingredients like lipids and flavours act as solvents of plastic materials, making them soft, even barrier plastic materials. Lipid and moisture content generally control the transfer of chemicals significantly. Storage temperature, storage duration, and contact surface are also important factors. A high storage temperature accelerates transfer while longer storage duration raises the amount of transfer.

In addition, nonvolatile migrants and flavours can be transferred when the food and packaging are in contact. In such cases, migration time is related to the cumulative duration of contact of migrants to food. Moreover, a package's entire surface area may affect the migration of volatiles and flavours. The use of barrier packaging or inert materials always reduces this type of transfer. Therefore, simple lamination or coating on packaging material surfaces is a very practical way to modify the barrier properties and, consequently, to control the chemical interactions between food and the package. Use of inert packaging materials with hermetic seals can also be used to contain volatile flavours.

NEW ADVANCES AND KEY AREAS OF CHANGE

SUSTAINABLE FOOD PACKAGING

Three concepts populate the key areas of change within food packaging. The first is the trend towards more sustainable packaging. While there are multiple definitions of sustainable packaging, the Sustainable Packaging Coalition, an international consortium of more than 200 industry members, offers the most accepted definition.

Sustainable packaging is characterized by the following criteria:

- It is beneficial, safe, and healthy for individuals and communities throughout its life cycle.
- It meets market criteria for performance and cost.
- It is sourced, manufactured, transported, and recycled using renewable energy.
- It maximizes the use of renewable or recycled source materials.
- It is manufactured using clean production technologies and best practices.
- It is made from materials healthy in all probable end-of-life scenarios.

- It is designed to optimize materials and energy.
- It is recovered effectively and used in biological and/or industrial cradle-to-cradle cycles.

Sustainability initiatives led by global legislation, retailers, and corporations guide package material choices, design, and food packaging sales for food packaging professionals. The revised 1997 European Commission's Packaging and Packaging Waste Directive, the 2007 REACH (Registration, Evaluation, and Authorization of Chemicals), and the BS EN 13432 standard (which defines compost-ability, degradability, and biodegradability) are examples of effective global legislative guides in a majority of the Group of 8 countries (Canada, France, Germany, Italy, Japan, Russia, United Kingdom, United States, and the European Union), BRIC (Brazil, Russia, India, and China), and the developing world.

In retail, IKEA and United Kingdom retailers have been long-term promoters of sustainable packaging and have launched impressive initiatives such as Plan A byMarks and Spencer, which defines food packaging material use. Also, Wal-Mart, the world's largest retailer, entered the sustainable packaging arena in 2006. Introduced in 2007, the Wal-Mart scorecard ranks packages compared with category competitors, based on environmental scores.

Finally, because a packaging material's source has been shown to be a defining factor in the final package's sustainability, corporations and environmental coalitions are working together to reduce the effect of packaging materials on global sources. International paper companies, the World Wildlife Fund, the Sustainable Forestry Initiative, and the Forest Stewardship Council verify the sources of wood for use in packaging.

Groups such as the Confederation of European Paper Industries (CEPI) in the European Union are addressing solutions to the remaining high environmental impact of paper, which is heavily dependent on water, gas, and energy. Other material-based working groups are taking similar actions.

USE OF PACKAGING SUPPLIER RELATIONSHIPS FOR COMPETITIVE ADVANTAGE

The second key area of change in food packaging is the use of packaging-development value chain relationships for competitive advantage. Suppliers to the food packaging industry are adding value in relationships within the traditional package development value chain. This chain extends from raw material generation, to conversion, to production and distribution, to retail, to consumer use and disposal.

But the formerly linear nature of this chain has become an integrated sphere that enables cross-fertilization of ideas fromvarious supply-chain functions directly to the food manufacturer. Investing in packaging supply chain relationships offers opportunities for focus, innovation, and technology transfer for a competitive advantage. For example, Starbucks engaged their packaging value chain—Mississippi River Corp., MeadWestvaco, and the Solo Cup Co.—to create an FDA-approved 10 per cent recycled-fibre coffee cup.

In addition, Green Mountain Coffee Roasters used its supply chain effectively when it introduced a compost able cup (the Ecotainer) with members of their supplier chain: DaniMer Scientific, NatureWorks, and International Paper. Also, Naturipe has used the value chain optimization to enable consistent packaging to stores from over 50 global locations.

EVOLUTION OF FOOD SERVICE PACKAGING

The third and final key area of change in food packaging is the evolution of food service packaging. Food service has grown to become a major part of consumer spending. As this trend increases, packaging plays a key role in ensuring food safety and providing convenience to consumers. For instance, proper package labeling allows food preparers to know the source of a food, its proper holding temperature, and the adequate cooking needed.

Ease of package opening allows for fewer utensils needed to open packages, which lessens contamination. While tracking and shelf-life extension technologies are employed in the food service industry to reduce the risk of food borne illness, proper heating and heat retention continue to be challenges. Consequently, packaging's role in ensuring proper heating and heat retention continues to increase. For example, CuliDish is a new product that uses varying levels of aluminum within a tray to package foods that require heating together with those that do not require heating.

The tray allows foods requiring high heat to be heated in the microwave at the same time with foods that do not require heat (such as salads). The food can be served in the same tray, thus reducing handling. More innovations in the areas of heat and heat retention are expected to assist in reducing food safety risks associated with improper cooking.

Two major convenience trends—meals eaten in transit and multi-component meals—have also advanced the food service packaging industry. The popularity of meals eaten in transit is evident since only 60 per cent of meals are prepared and eaten at home and 20 per cent of consumers eat on their way to another location instead of in a fixed position. On-the-go food consumption has resulted in packaging that contains a greater variety of foods.

Technologies that support this trend include edible films to enrobe food particles and edible wraps that peel off, allowing consumers to eat numerous foods while in transit. Food package design innovations such as modular folding cartons with flip-off lids, pouches that are easily opened or have a seatbelt flap to hold food close to consumers, and reusable packaging have contributed to the increase of eating while on the go. The growth of multi-component meals in food service stems from multiple types of foods being ordered at quick-service restaurants. This innovation allows for ease of food service preparation and less waste. Multi-component packaging also provides consumers with a presentation platform that makes multi-component foods easy to consume. Examples include reusable trays that clip together into 1 larger serving tray such as that used in the Les Petits Grande line of products; KFC's triple dip strip cartons, which allow consumers to select from 1 of 3 dips when dipping chicken strips; and the folding tray that allows consumers to carry 2 cups of coffee with 1 hand.

ADVANCES IN FOOD PACKAGING DISTRIBUTION

RFID SYSTEMS FOR PACKAGED FOODS: ARCHITECTURE AND WORKING PRINCIPLES

Radio frequency identification (RFID) is a system that uses radio waves to track items wirelessly. RFID makes use of tags or transponders (data carriers), readers (receivers), and computer systems (software, hardware, networking, and database). The tags consist of an integrated circuit, a tag antenna, and a battery if the tag is passive (most active tags do not require battery power).

The integrated circuit contains a non-volatile memory microchip for data storage, an AC/DC converter, encode/decode modulators, a logic control, and antenna connectors. The wireless data transfer between a transponder/tag and a reader makes RFID technology far more flexible than other contact identifications, such as the barcode system and thus makes it ideal for food packaging.

The working principles of an RFID system are as follows:

- Data stored in tags are activated by readers when the objects with embedded tags enter the electromagnetic zone of a reader;
- Data are transmitted to a reader for decoding; and
- Decoded data are transferred to a computer system for further processing.

Tag frequency is related to the working principles of an RFID system (for example, magnetic coupling or electric coupling) and the reading range. Frequency depends on the type of tag, reader, and cost. The typical RFID frequencies are low frequency, high frequency, ultra high frequency, and microwave frequency. Generally, low frequency systems have short reading ranges, slow read speeds, and lower cost while higher frequency RFID systems are utilized when longer read ranges and fast reading speeds are required. Microwave frequency requires active RFID tags.

ELECTRONIC PRODUCT CODES

In 2004, Electronic Product Code (EPC) Global Network began developing a second generation RFID protocol: EPC Class 1 version 2. The main goal of Gen 2 is to create a single global standard that is compatible with ISO standards. The tags would work in various countries that have different commercial band frequency. EPC is the most significant function of RFID contributing to commercial industry.

EPC improves the traceability of items and facilitates efficient product recall and authenticity. EPC is similar to Universal Product Code (UPC), which is commonly used in bar codes. Compared to UPC, which uses 12 digits of numbers, EPC has 64 to 256 bits of alphanumeric data. The most common EPC has 96 bits. The first obstacle for wide utilization of RFID is the cost for tags. Tags are still too expensive for use on individual primary packages.

The infrastructure required for RFID systems (including readers, database servers with communication systems, and other information technology to process huge amounts of data) is costly and needs to be shared with all users in supply chains. The global use of EPC also requires compatibility among various regulations and standards of radio frequency.

The biggest hurdle to wide utilization of EPC for RFID is the potential problems of privacy protection. A hidden reading system could collect all data from tags of items and also RFID card holders for the purpose of data stealing or data removing. Guidelines for the ethical use of RFID systems for data collecting, data handling, and system security need to be established.

RFID FOR THE FOOD INDUSTRY

RFID has recently found its way into numerous applications in the food industry, ranging from food monitoring and traceability to enhancing food safety, to improving supply chain efficiency. The major benefits of RFID technology in the food industry are greater speed and efficiency in stock rotation and better tracking of products throughout the chain, resulting in improved on-shelf availability at the retail level and enhanced forecasting. The technology is well suited for many operations in food manufacturing and supply chain management.

An RFID-based resource management system can help users handle warehouse operating orders by retrieving and analysing warehouse data, which could save time and cost. The use of RFID in the food industry is currently focused on tracking and identification. When RFID technology becomes more established in the food industry, the integration of food science knowledge will be necessary to develop the intelligent food packaging application for food quality and safety.

Some food companies have already integrated RFID into manufacturing and distribution. Retail chains such as Wal-Mart and Home Depot have been testing the technology for distribution. In 2003, Wal-Mart issued a mandate requiring its top 100 suppliers to use RFID tags on all cases and pallets entering its distribution centers by 2005.

RFID compliance is a long-term project for Wal-Mart; more of its suppliers are expected to be compliant by the end of 2008. Other major players advocating RFID technology are the U.S. Department of Defence and major retailers such as Albertsons, Target, Tesco, and Marks and Spencer. RFID technology also provides security and safety benefits for food companies through tracking the origin of supplies. For example, a small California winery uses RFID to track its barrels and to enhance wine making by streamlining data collection.

The company planted RFID tags on tanks and harvesting bins, allowing better control of wine production and tracking. In addition, by attaching an RFID tag to a package, the package becomes intelligent because the stored data provide valuable information that can be stored and read by appliances. This intelligent packaging technology is also being extended to refrigeration and freezing. Appliances can communicate with the packages and identify information related to the storage of the packaged products. Despite these benefits, other factors such as cost of the technology and recycling ability need to be considered.

SUPPLY CHAIN MANAGEMENT, TRACEABILITY, AND RECALL

As a tool for tracking, RFID is a promising technology to food supply chain management. The FDA, 1307 recalls of processed foods occurred between 1999 and 2003; these recalls could have been avoided with a technology such as RFID. Exposure to risk at any one of the stages of the processed food supply chain would result in a domino-effect breakdown that could affect the smooth running of an entire supply chain.

All stages of the supply chain (from farming, processing, transportation, manufacturing, retailing, and warehousing to consumption) are equally critical to food recall problems. If RFID technology were combined with Hazard Analysis and Critical Control Point systems, the supply chain stages would be integrated, traceable, and effectively managed by food processors to reduce the number of recalls significantly. The outstanding tracing abilities of RFID tags to individual food product could enable manufacturers to audit every single phase of a product in a retail unit, monitoring correct handling, transportation, storage, and delivery.

NEW TAG FOR USE ON METAL AND METAL PACKAGING AND HIGH WATER CONTENT PRODUCTS

The RFID system is not infallible; it has some weaknesses, such as the shielding effect of metal, which affects signal transduction. Data cannot be read correctly when tags are attached to metal on the surface or inside the package. Until recently, high frequency RFID tags were used on metal beer barrels. However, low frequency RFID (125 kHz to 135 kHz) has less loss of radio signal by the metal materials in the high magnetic field of a reader compared to higher frequency signals. Another issue is that water molecules can absorb microwave signals, resulting in signal loss or interference during data acquisition from microwave RFID tags. Since most foods contain high moisture, this signal interference requires further study to enable application of the technology in the food industry. It is worth noting that low, high, and ultra-high frequency (UHF) tags can be used in high water systems since water does not interfere with their signal; on the contrary, ice absorbs UHF radio signals.

For example, a product such as ice cream (which is technically defined as partly frozen foam with ice crystals and air bubbles occupying a majority of the space) contains about 72 per cent frozen water and thus interferes with UHF radio signals. Ice cream manufacturers have overcome this product interference by placing tags over an air gap in the containers. Companies with diverse product lines and high levels of automation will likely encounter significant technical barriers to RFID implementation.

NANOCOMPOSITES AND OTHER EMERGING NANOTECHNOLOGIES IN FOOD PACKAGING

Nanotechnology has the potential to transformfood packaging materials in the future. Such nanoscale innovation could potentially introduce many amazing new improvements to food packaging in the forms of barrier and mechanical properties, detection of pathogens, and smart and active packaging with food safety and quality benefits. The nanolayer of aluminum that coats the interior of many snack food packages is one common example of the role that nanotechnology already plays in food packaging. The market for nanotechnology in food packaging in 2006 was estimated at $66 million and is expected to reach $360 million in 2008.

Nanomaterials are abundant in nature and numerous techniques are available to fabricate various nanomaterials. Nanoparticles can be produced top down from larger structures by grinding, use of lasers, and vaporization followed by cooling. Alternately, bottom-up methods are commonly used for synthesis of complex nanoparticles. These methods include solvent extraction/ evaporation, crystallization, self-assembly, layer-by-layer deposition, microbial synthesis, and biomass reactions. All of these are being researched for potential application in food packages in the future. One group of nanomaterials at the forefront of food packaging development is nanocomposites.

NANOCOMPOSITES

Nanocomposite packages are predicted to make up a significant portion of the food packaging market in the near future. Principia Markets, a consulting firmthat tracks the plastics market, estimates that the market for nanocomposites will reach 1 billion pounds by 2010. Many nanocomposite food packages are either already in the marketplace or being developed. The majority of these are targeted for beverage packaging. In large part, the impetus for this predicted growth is the extraordinary benefits nanoscience offers to improve food packages.

Improvements in fundamental characteristics of food packaging materials such as strength, barrier properties, antimicrobial properties, and stability to heat and cold are being achieved using nanocomposite materials. In the late 1980s, Toyota was the 1st company to commercialize nanocomposite materials. They found that the addition of 5 per cent-by-weight nano-sized montmorillonite clay significantly increased the mechanical and thermal properties of different grades of nylon. Nanocomposite materials are now used in gasoline tanks, bumpers, and interior and exterior panels. Research on use of nanocomposites for food packaging began in the 1990s.

Most of the research has involved the use of montmorillonite clay as the nanocomponent in a wide range of polymers such as polyethylene, nylon, polyvinyl chloride, and starch. Amounts of nanoclays incorporated vary from 1 per cent to 5 per cent by weight. Nanocomponents must have 1 dimension less than 1 nm wide. The lateral dimensions, on the other hand, can be as large as several micrometers, leading to high aspect ratios (ratio of length to thickness) of many of these materials.

The high surface area results in unique properties when nanocomposites are incorporated into packages. There are 3 common methods used to process nanocomposites: solution method, *in situ* or interlamellar polymerization technique, and melt processing. The solution method can be used to form both intercalated and exfoliated nanocomposite materials. In the solution method, the nanocomposite clay is first swollen in a solvent. Next, it is added to a polymer solution, and polymer molecules are allowed to extend between the layers of filler.

The solvent is then allowed to evaporate. The *in situ* or interlamellar method swells the fillers by absorption of a liquid monomer. After the monomer has penetrated in between the layers of silicates, polymerization is initiated by heat, radiation, or incorporation of an initiator. The melt method is the most commonly used method due to the lack of solvents. In melt processing, the nanocomposite filler is incorporated into a molten polymer and then formed into the final material.

Generally, there are 3 possible arrangements for layered silicate clay nanocomposite materials:

1. Nonintercalated,
2. Intercalated, and
3. Exfoliated or delaminated.

In nonintercalated materials the polymer does not fit between the layered clay, leading to a microphase separated final structure. In intercalated systems, the polymer is located between clay layers, increasing interlayer spacing. Some degree of order is retained in parallel clay layers, which are separated by alternating polymer layers with a repeated distance every few nanometers.

Exfoliated systems achieve complete separation of clay platelets in random arrangements. This is the ideal nanocomposite arrangement but is hard to achieve. Bayer produces transparent nanocomposite plastic films and coatings called Durethan, which contains clay nanoparticles dispersed throughout the plastic. Large amounts of silicate nanoparticles are interspersed in polyamide films. These nanoparticles block oxygen, carbon dioxide, and moisture from reaching fresh meats and others foods. The nanoclay particles act as impermeable obstacles in the path of the diffusion process, thereby extending the shelf life of foods while improving their quality.

The final package is also considerably lighter, stronger, and more heat-resistant. In years past, packaging beer in plastic bottles was not possible due to oxidation and flavour problems. Recently, however, this challenge has been overcome using nanotechnology. For example, Nanocor, a subsidiary of Amcol International Corp., is producing nanocomposites for use in plastic beer bottles that facilitate a 6-mo shelf life. By combining the nanocomposite and oxygen scavenger technologies, a new family of barrier nylons was recently developed for use in multilayer, co-injection blow-molded PET bottles.

In the near future, nanocrystals embedded in plastic bottles may increase beer shelf life up to 18 mo by minimizing loss of carbon dioxide from and entrance of oxygen into bottles. Similar materials are being developed to extend the shelf life of soft drinks. Another advantage of these nanocomposite bottles is that theirweight is considerably less, thereby reducing transportation costs. A considerable amount of research is also occurring in the area of biodegradable nanocomposite food packages.

By pumping carbohydrates and clay fillers through high shear cells, films can be produced with exfoliated clay layers. These films act as very effective moisture barriers by increasing the tortuosity of the path water must take to penetrate the films. Significant increases in film strength are also frequently achieved in these types of materials. Starch and chitosan are two of the most studied biodegradable matrices. In the future, these types of biodegradable nanocomposite food packages may be found in the marketplace.

OTHER NANOTECHNOLOGIES

Carbon nanotubes are cylinders with nanoscale diameters that can be used in food packaging to improve its mechanical properties. In addition, it was recently discovered that they may also exert powerful antimicrobial effects. *Escherichia coli* died immediately upon direct contact with aggregates of carbon nanotubes. Presumably, the long, thin nanotubes punctured the *E. coli* cells, causing cellular damage. Single-walled carbon nanotubes may eventually serve as building blocks for antimicrobial materials. Nano-wheels were also recently developed to improve food packaging.

Inorganic alumina platelets have been self-assembled into wagon-wheel shaped structures that are incorporated into plastics to improve their barrier and mechanical properties. This was the first time large wheel-shaped molecules had been formed. The addition of nanosensors to food packages is also anticipated in the future. Nanosensors could be used to detect chemicals, pathogens, and toxins in foods.

Numerous research reports describe detection methods for bacteria, viruses, toxins, and allergens using nanotechnology. For example, adhering antibodies to *Staphylococcus* enterotoxin B onto poly(dimethyl-siloxane) chips formed biosensors that have a detection limit of 0.5 ng/mL. Nanovesicles have been developed to simultaneously detect *E. coli* 0157:H7, *Salmonella* spp., and *Listeria monocytogenes*. Liposome nanovesicles have been devised to detect peanut allergen proteins.

In addition, AgroMicron has developed a NanoBioluminescence detection spray containing a luminescent protein that has been engineered to bind to the surface of microbes such as *Salmonella* and *E. coli*. When bound, it emits a visible glow that varies in intensity according to the amount of bacterial contamination. This product is being marketed under the name BioMark. Nanosensors Inc. is another company pursuing this potential. Through a license agreement with Michigan State University, a nanoporous silicon-based biosensor has been developed to detect *Salmonella* and *E. coli*.

A prototype nanobiosensor was recently tested to detect *Bacillus cereus* and *E. coli* and was found to be able to detect multiple pathogens faster and more accurately than current devices. Finally, Mahadevan Iyer and his colleagues at Georgia Institute of Technology are experimenting with integrating nanocomponents in ultra-thin polymer substrates for RFID chips containing biosensors that can detect foodborne pathogens or sense the temperature or moisture of a product. DNA biochips are already under development to detect pathogens.

Researchers at the Univ. of Pennsylvania and Monell Chemical Sciences Center have used nano-sized carbon tubes coated with strands of DNA to create nanosensors with abilities to detect odours and tastes. A single strand of DNA serves as the sensor and a carbon nanotube functions as the transmitter. Using similar technologies, electronic tongue nanosensors are being developed to detect substances in parts per trillion, which could be used to trigger colour changes in food packages to alert consumers when food is spoiled.

A unique aspect of these biochips is that the DNA is self-assembled onto the chips and repairs itself if damaged. In addition, researchers at Cornell Univ. have invented synthetic DNA barcodes to tag pathogens and monitor pathogens. The nanobarcodes fluoresce under ultraviolet light when target compounds are detected. Another colour-changing film that could find its way into food packages is polymer opal films. Scientists at the United Kingdom's Univ. of Southampton and the Deutsches Kunststoff Inst. in Germany developed these unique self-assembled structures from arrays of spheres stacked in 3 dimensions. Polymer opal films belong to a class of materials known as photonic crystals. The crystals are built of tiny repeating units of carbon nanoparticles wedged between spheres, leading to intense colors that mimic the colors associated with the photonic crystals found on butterfly wings and peacock feathers. Photonic crystals could be used to produce unique food packaging materials that change colour.

CONCLUSION

The food industry has seen great advances in the packaging sector since its inception in the 18th century with most active and intelligent innovations occurring during the past century. These advances have led to improved food quality and safety. While some innovations have stemmed from unexpected sources, most have been driven by changing consumer preferences. The new advances have mostly focused on delaying oxidation and controlling moisture migration, microbial growth, respiration rates, and volatile flavours and aromas.

This focus parallels that of food packaging distribution, which has driven change in the key areas of sustainable packaging, use of the packaging value chain relationships for competitive advantage, and the evolving role of food service packaging. Nanotechnology has potential to influence the packaging sector greatly. Nanoscale innovations in the forms of pathogen detection, active packaging, and barrier formation are poised to elevate food packaging to new heights.

3

Dairy Process Biotechnology

INTRODUCTION

Food processing makes use of various unit operations and technologies to convert relatively bulky, perishable and typically inedible raw materials into more useful shelf-stable and palatable foods or potable beverages. Processing contributes to food security by minimizing waste and losses in the food chain and by increasing food availability and marketability. Food is also processed in order to improve its quality and safety. Food safety is a scientific discipline that provides assurance that food will not cause harm to the consumer when it is prepared and/or eaten according to its intended use.

Biotechnology as applied to food processing in most developing countries makes use of microbial inoculants to enhance properties such as the taste, aroma, shelf-life, texture and nutritional value of foods. The process whereby micro-organisms and their enzymes bring about these desirable changes in food materials is known as fermentation.

Fermentation processing is also widely applied in the production of microbial cultures, enzymes, flavours, fragrances, food additives and a range of other high value-added products. These high value products are increasingly produced in more technologically advanced developing countries for use in their food and non-food processing applications. Many of these high value products are also imported by developing countries for use in their food-processing applications.

This document will discuss the prospects and potential of applying biotechnology in food processing operations and to address safety issues in food systems with the objective of addressing food security and responding to changing consumer trends in developing countries. It is important to note that food safety evaluation or risk assessment will not be discussed here. Instead, this chapter will focus on the context of biotechnologies as applied to food safety.

Technologies applied in the processing of food must assure the quality and safety of the final product. Safe food is food in which physical, chemical or microbiological hazards are present at a level that does not present a public health risk. Safe food can, therefore, be consumed with the assurance that there are no serious health implications for the consumer. Recent food scares such as mad cow disease and the melamine contamination of food products have increased consumer concern for food safety. As incomes rise, consumers are increasingly willing to pay a premium for quality, safety and convenience.

A range of technologies is applied at different levels and scales of operation in food processing across the developing world. Conventional or "low-input" food processing technologies include drying, fermentation, salting, and various forms of cooking, including roasting, frying, smoking, steaming, and oven baking. Low-income economies are likely to employ these as predominant technologies for the processing of staple foods.

Many of these technologies make use of a simple, often rudimentary, technological base. Medium levels of processing technologies such as canning, oven drying, spray drying, freeze drying, freezing, pasteurization, vacuum packing, osmotic dehydration and sugar crystallization are widely applied in middle-and upper middle-income economies.

Higher-level, more capital-intensive food-processing technologies such as high-temperature short-time pasteurization and high-pressure low-temperature food processing are widely employed in middle-and upper middle-income economies. Functional additives and ingredients produced using fermentation processes are generally incorporated into food-processing operations that make use of higher-level technologies.

Traditional methods of food-safety monitoring such as the detection of pathogenic bacteria in food are generally based on the use of culture media. These are the techniques of choice in low-and lower-middle-income economies which lack the resources, infrastructure and technical capacity to utilize modern biotechnological techniques.

Conventional bacterial detection methods are time-consuming multi-step procedures. At least two to three days are required for the initial isolation of an organism, followed by the requirement for several days of additional confirmatory testing. Biotechnology-based methods can provide accurate results within a relatively short time frame.

Biotechnological developments have resulted in the widespread availability of low-cost rapid methods of identification when compared with the significant cost/time requirements of conventional techniques. Lower-middle-income economies apply both traditional and more sophisticated methods for monitoring the microbiological quality of foods and their conformance to international standards.

A number of case studies are described in the text to demonstrate the utility of biotechnology-based applications in food processing and food safety. These case studies provide the basis for the development of strategic interventions designed to upgrade food processing and food safety in developing countries through the application of biotechnology.

STOCKTAKING: LEARNING FROM THE PAST

BIOTECHNOLOGY: DEFINITION AND SCOPE

For the purpose of this chapter, biotechnology is defined in accordance with the Convention on Biological Diversity, *i.e. "any technological application that uses biological systems, living organisms, or derivatives thereof, to make or modify products or processes for specific use"*.

Biotechnology in the food processing sector makes use of micro-organisms for the preservation of food and for the production of a range of value-added products such as enzymes, flavour compounds, vitamins, microbial cultures and food ingredients. Biotechnology applications in the food-processing sector, therefore, target the selection and manipulation of micro-organisms with the objective of improving process control, product quality, safety, consistency and yield, while increasing process efficiency.

Biotechnological processes applicable to the improvement of microbial cultures for use in food-processing applications include traditional methods of genetic improvement such as classical mutagenesis and conjugation. These methods generally focus on improving the quality of micro-organisms and the yields of metabolites. Hybridization is also used for the improvement of yeasts involved in baking, brewing and in beverage production. *Saccharomyces cerevisiae* strains have, for example, been researched for improved fermentation, processing and biopreservation abilities, and for capacities to increase the wholesomeness and sensory quality of wine.

Recombinant gene technology, the best-known modern biotechnology, is widely employed in research and development for strain improvement. The availability of genetic manipulation tools and the opportunities that exist to improve the microbial cultures associated with food fermentations are tempered by concerns over regulatory issues and consumer perceptions. *Genetically modified* (GM) microbial cultures are, however, used in the production of enzymes and various food-processing ingredients such as monosodium glutamate, polyunsaturated fatty acids and amino acids. Biotechnology is also widely employed as a tool in diagnostics in order to monitor food safety, prevent and diagnose food-borne illnesses and verify the origins of foods. Techniques applied in the assurance of food safety focus on the detection and monitoring of hazards whether biological, chemical or physical. These applications will be explored and discussed in subsequent sections.

CURRENT STATUS OF THE APPLICATION OF TRADITIONAL AND NEW BIOTECHNOLOGIES IN FOOD PROCESSING IN DEVELOPING COUNTRIES

Methods of Microbial Inoculation in Food Fermentations

The fermentation bioprocess is the major biotechnological application in food processing. It is often one step in a sequence of food-processing operations, which may include cleaning, size reduction, soaking and cooking. Fermentation bioprocessing makes use of microbial inoculants for enhancing properties such as the taste, aroma, shelf-life, safety, texture and nutritional value of foods.

Microbes associated with the raw food material and the processing environment serve as inoculants in spontaneous fermentations, while inoculants containing high concentrations of live micro-organisms, referred to as starter cultures, are used to initiate and accelerate the rate of fermentation processes in non-spontaneous or controlled fermentation processes. Microbial starter cultures vary widely in quality and purity.

Starter culture development and improvement is the subject of much research both in developed and in developing countries. While considerable work on GM starter culture development is ongoing at the laboratory level in developed countries, relatively few GM micro-organisms have been permitted in the food and beverage industry globally. In 1990, the United Kingdom became the first country to permit the use of a live genetically modified organism (GMO) in food. It was a baker's yeast, engineered to improve the rate at which bread dough rises by increasing the efficiency with which maltose is broken down.

This modification was done by using genes from yeast and placing them under a strong constitutive promoter. The United Kingdom has also approved a GM brewer's yeast for beer production. By introducing a gene encoding glucoamylase from yeast, better utilization of carbohydrate present in conventional feedstock can be obtained, resulting in increased yields of alcohol and the ability to produce a full-strength, low-carbohydrate beer. More recently, two genetically modified yeast strains were authorized for use in the North American wine industry.

Current literature documents volumes of research reports on the characterization of microbes associated with the production of traditional fermented foods in developing countries. Relatively few of these studies document the application of the diagnostic tools of modern biotechnology in developing and designing starter cultures.

The development and improvement of microbial starters has been a driving force for the transformation of traditional food fermentations in developing countries from an "art" to a science. Microbial starter culture development has also been a driving force for innovation in the design of equipment suited to the hygienic processing of traditional fermented foods under controlled conditions in many developing countries.

Starter culture improvement, together with the improvement and development of bioreactor technology for the control of fermentation processes in developed countries, has played a pivotal role in the production of high-value products such as enzymes, microbial cultures, and functional food ingredients. These products are increasingly produced in more advanced developing economies, and are increasingly imported by less advanced developing countries, as inputs for their food processing operations.

Spontaneous Inoculation of Fermentation Processes

In many developing countries, fermented foods are produced primarily at the household and village level, using spontaneous methods of inoculation. Spontaneous fermentations are largely uncontrolled. A natural selection process, however, evolves in many of these processes which eventually results in the predominance of a particular type or group of micro-organisms in the fermentation medium.

A majority of African food-fermentation processes make use of spontaneous inoculation. Major limitations of spontaneous fermentation processes include their inefficiency, low yields of product and variable product quality. While spontaneous fermentations generally enhance the safety of foods owing to a reduction of pH, and through detoxification, in some cases there are safety concerns relating to the bacterial pathogens associated with the raw material or unhygienic practices during processing.

"Appropriate" Starter Cultures as Inoculants of Fermentation Processes

"Appropriate" starter cultures are widely applied as inoculants across the fermented food sector, from the household to industrial level in low-income and lower-middle-income economies. These starter cultures are generally produced using a backslopping process which makes use of samples of a previous batch of a fermented product as inoculants.

Appropriate starter cultures are widely applied in the production of fermented fish sauces and fermented vegetables in Asia and in cereal or grain fermentations in African and Latin American countries. The inoculation belt used in traditional fermentations in West Africa serves as a carrier of undefined fermenting micro-organisms, and is one example of an appropriate starter culture.

It generally consists of a woven fibre or mat or a piece of wood or woven sponge, saturated with "high"-quality product of a previously fermented batch. It is immersed into a new batch, in order to serve as an inoculant. The inoculation belt is used in the production of the indigenous fermented porridges, "uji" and "mawe", as well as in the production of the Ghanaian beer, "pito".

Iku, also referred to as iru, is yet another example of an "appropriate" starter culture produced by backslopping. This starter culture is produced from concentrated fermented dawadawa, mixed with ground unfermented legumes, vegetables such as pepper, and cereals, such as ground maize. It is stored in a dried form and is used as an inoculant in dawadawa fermentations in West Africa. A range of appropriate starter cultures, either in a granular form or in the form of a pressed cake is used across Asian countries as fermentation inoculants. These traditional mould starters are generally referred to by various names such as *marcha* or *murcha* in India, *ragi* in Indonesia, *bubod* in the Philippines, *nuruk* in Korea, *koji* in Japan, *ragi* in Malaysia and *Loog-pang* in Thailand. They generally consist of a mixture of moulds grown under non-sterile conditions.

Defined Starter Cultures as Inoculants of Fermentation Processes

Few defined starter cultures have been developed for use as inoculants in commercial fermentation processes in developing countries. Nevertheless, the past ten years have witnessed the development and application of laboratory-selected and pre-cultured starter cultures in food fermentations in a few developing countries. These developments have taken place primarily in Asian countries.

"Defined starter cultures" consist of single or mixed strains of micro-organisms. They may incorporate adjunct culture preparations that serve a food-safety and preservative function. Adjunct cultures do not necessarily produce fermentation acids or modify texture or flavour, but are included in the defined culture owing to their ability to inhibit pathogenic or spoilage organisms. Their inhibitory activity is due to the production of one or several substances such as hydrogen peroxide, organic acids, diacetyl and bacteriocins.

By and large, defined cultures are produced by pure culture maintenance and propagation under aseptic conditions. They are generally marketed in a liquid or powdered form or else as a pressed cake. Loog-pang, a defined culture marketed in Thailand in the form of a pressed rice cake, consists of *Saccharomyces cerevisiae, Aspergillus oryzae* or *Rhizopus* sp. and Mucor. Loog-pang has a shelf life of 2–3 days at ambient temperature and 5–7 days under refrigerated conditions.

Ragi cultures are commercially produced by the Malaysian Agricultural Research and Development Institute by mixing a culture inoculum which generally consists of *Rhizopus Oligosporus* with moistened sterile rice flour, and incubating it at ambient temperature for four days. This starter has a shelf life of two weeks under refrigerated conditions. It is widely used as an inoculant in the production of traditional Malaysian fermented foods. Ragi-type starter cultures for the production of a range of fermented Indonesian products such as oncom, tape and tempeh are currently marketed via the internet.

Defined starter cultures are also widely imported by developing countries for use in the commercial production of dairy products such as yogurt, kefir, cheeses and alcoholic beverages. Many of these cultures are tailored to produce specific textures and flavours. In response to growing consumer interest in attaining wellness through diet, many yogurt cultures also include probiotic strains. Probiotics are currently produced in India for use as food additives, dietary supplements and for use in animal feed. Methodologies used in the development and tailoring of these starters are largely proprietary to the suppliers of these starters. Monosodium glutamate and lactic acid, both of which are used as ingredients in the food industry, are produced in less-advanced developing countries using defined starter cultures.

Defined Starter Cultures Developed Using the Diagnostic Tools of Advanced Biotechnologies

The use of DNA-based diagnostic techniques for strain differentiation can allow for the tailoring of starter cultures to yield products with specific flavours and/or textures. *Random amplified polymorphic DNA* (RAPD) techniques have been applied in, for example, Thailand, in the molecular typing of bacterial strains and correlating the findings of these studies to flavour development during the production of the fermented pork sausage, nham. The results of these analyses led to the development of three different defined starter cultures which are currently used for the commercial production of products having different flavour characteristics.

GM Starter Cultures

To date, no commercial GM micro-organisms that would be consumed as living organisms exist. Products of industrial GM producer organisms are, however, widely used in food processing and no major safety concerns have been raised against them. Rennet which is widely used as a starter in cheese production across the globe is produced using GM bacteria. Thailand currently makes use of GM *Escherichia coli* as an inoculant in lysine production.

Many industrially important enzymes such as α-amylase, gluco-amylase, lipase and pectinase and bio-based fine chemicals, such as lactic acid, amino acids, antibiotics, nucleic acid and polysaccharides, are produced in China using GM starter cultures. Other developing countries which currently produce enzymes using recombinant micro-organisms include Cuba, Brazil, India, and Argentina.

Food Additives and Processing Aids

Enzymes, amino acids, vitamins, organic acids, polyunsaturated fatty acids and certain complex carbohydrates and flavouring agents used in food formulations are currently produced using GM micro-organisms.

Enzymes

Enzymes occur in all living organisms and catalyze biochemical reactions that are necessary to support life. They are commonly used in food processing and in the production of food ingredients. The use of recombinant DNA technology has made it possible to manufacture novel enzymes that are tailored to specific food processing conditions. Alpha amylases with increased heat stability have, for example, been engineered for use in the production of high-fructose corn syrups.

These improvements were accomplished by introducing changes in the α-amylase amino acid sequences through DNA sequence modifications of the α-amylase genes. Bovine chymosin used in cheese manufacture was the first recombinant enzyme approved for used in food by the US Food and Drug Administration. The Phospholipase A1 gene from *Fusarium venenatum* is expressed in GM *Aspergillus oryzae* to produce the phospholipase A1 enzyme used in the dairy industry for cheese manufacture to improve process efficiencies and cheese yields.

Considerable progress has been made in recent times towards the improvement of microbial strains used in the production of enzymes. Microbial host strains developed for enzyme production have been engineered to increase enzyme yields by deleting native genes encoding extracellular proteases. Certain fungal producing strains have also been modified to reduce or eliminate their potential for producing toxic metabolites.

Enzymes used in food processing have historically been considered non-toxic. Some characteristics arising from their chemical nature and source, such as allergenicity, activity-related toxicity, residual microbiological activity and chemical toxicity are, however of concern. These attributes of concern must, however, be addressed in light of the growing complexity and sophistication of the methodologies used in the production of food-grade enzymes.

Safety evaluation of all food enzymes, including those produced by GM micro-organisms, is essential if consumer safety is to be assured. Enzymes produced using GM micro-organisms wherein the enzyme is not part of the final food product have specifically been evaluated by the *Joint FAO/WHO Expert Committee on Food Additives* (JECFA). Safety evaluations have been conducted using the general specifications and considerations for enzyme preparations used in food processing. Preparations of asparaginase enzymes have also been evaluated by JECFA.

Flavours, Amino Acids and Sweeteners

Volatile organic chemicals such as flavours and aromas are the sensory principles of many consumer products and govern their acceptance and market success. Flavours produced using micro-organisms currently compete with those from traditional agricultural sources. Berger more than 100 commercial aroma chemicals are derived using biotechnology either through the screening for overproducers, the elucidation of metabolic pathways and precursors or through the application of conventional bioengineering.

Recombinant DNA technologies have also enhanced efficiency in the production of non-nutritive sweeteners such as aspartame and thaumatin. Market development has been particularly dynamic for the flavour enhancer glutamate which is produced by the fermentation of sugar sources such as molasses, sucrose or glucose using high-performance strains of *Corynebacterium glutamicum* and *Escherichia coli*. Amino acids produced through biotechnological processes are also of great interest as building blocks for active ingredients used in a variety of industrial processes.

Current Status of the Application of Traditional and New Biotechnologies in Food Safety and in Quality Improvement in Developing Countries

Food Safety Issues and Concerns in Food Fermentation Processing

Microbial activity plays a central role in food fermentation processes, resulting in desirable properties such as improvements in shelf-life and quality attributes such as texture and flavour. Pathogenic organisms are, however, of prime concern in fermented foods. Anti-nutritional factors such as phytates, tannins, protein inhibitors, lectins, saponins, oligosaccharides and cyanogenic glucosides are naturally occurring components of raw materials commonly used in food fermentations in developing countries.

Contamination of the fermentation process can pose a major health risk in the final fermented product. Methodologies for identifying and monitoring the presence of chemical and biochemical hazards in fermented foods are, therefore, a critical need. Furthermore, with growing consumer interest in the credence attributes of the products that they consume, and the premium currently being placed on quality linked to geographical origin, the traceability of foods with selected properties is of increasing importance.

Advances in Microbial Genetics

In recent times, the genetic characterization of micro-organisms has advanced at a rapid pace with exponential growth in the collection of genome sequence information, high-throughput analysis of expressed products *i.e.*, transcripts and proteins and the application of bioinformatics which allows high throughput comparative genomic approaches that provide insights for further functional studies.

Genome sequence information, coupled with the support of highly advanced molecular techniques, have allowed scientists to establish mechanisms of various host-defensive pathogen counter-defensive strategies and have provided industry with tools for developing strategies to design healthy and safe food by optimizing the effect of probiotic bacteria, the design of starter culture bacteria and functional properties for use in food processing.

Characterization of the genomes of lactic acid probiotics has, for example, shed light on the interaction of pathogens with lactic acid bacteria. Nucleotide sequences of the genomes of many important food microbes have recently become available. *Saccharomyces cerevisiae* was the first food microbe for which a complete genome sequence was characterized. This was followed by genome sequencing of the related yeast, *Kluyveromyces lactis* as well as filamentous fungi which are major enzyme producers and have significant applications in the food-processing industry.

Genome nucleotide sequences of many Gram-positive bacteria species have also been completed. The *Bacillus subtilis* genome was the first to be completed, followed by that of the *Lactococcus lactis* genome. Genome sequences of food-borne pathogens such as *Campylobacter jejuni*, verocytotoxigenic *Escherichia coli* O157:H7 and *Staphylococcus aureus* have also been completed. Genome sequences of microbes that are of importance in food processing, such as *Lactobacillus plantarum* are also available. The genome of *Clostridium botulinum*, responsible for food poisoning, was also recently completed.

Detection of Pathogens

The rapid detection of pathogens and other microbial contaminants in food is critical to assess the safety of food products. Traditional methods to detect food-borne bacteria often rely on time-consuming growth in culture media, followed by isolation, biochemical identification, and sometimes serology. Recent technological advances have improved the efficiency, specificity and sensitivity of detecting micro-organisms. Detection technologies employ the *polymerase chain reaction* (PCR) assay.

Short fragments of DNA (probes) or primers are hybridized to a specific sequence or template, which is subsequently enzymatically amplified by the Taq polymerase enzyme using a thermocycler. In theory, a single copy of DNA can be amplified a million-fold in less than 2 hours with the use of PCR techniques; hence, the potential of PCR to eliminate or greatly reduce the need for cultural enrichment.

The genetic characterization of genome sequence information has further facilitated the identification of virulence nucleotide sequences for use as molecular markers in pathogen detection. Multiplex real-time PCR methods are now available to identify the *E. coli* O157:H7 serogroup. PCR-based identification methods are also available for *Vibrio cholerae* and for major food-related microbes such as *Campylobacter jejuni, C. coli, Yersinia enterocolitica, Hepatitis A virus, Salmonella, Staphylococcus aureus*.

Sophisticated cultural media such as chromogenic or fluorogenic media are not readily used in low-income economies but are relatively widespread in lower-middle-income and upper-middle-income economies. The use of immunoassays such as enzyme-linked immunosorbent assay (ELISA) is also very limited in low-income economies but is more widespread in the form of diagnostic kits in lower-middle and upper-middle-income economies.

DNA methods, which require elaborate infrastructure and high technical competence, find minimal application in lower-income and some lower-middle-income economies. There are movements towards implementing safety-control programmes such as the application of Hazard Analysis and Critical Control Point (HACCP) in food fermentations in many developing countries.

The application of HACCP necessitates the deployment of good agricultural practices, good manufacturing practices (GMPs) and good hygienic practices (GHPs) and the monitoring of critical control points for potential microbial and chemical contamination during bioprocessing. Rigorous adherence to sanitary practices in the processing environment necessitates rapid, dynamic, sensitive, specific as well as versatile and cost-effective assay methods. The molecular approach of biotechnology entails near-time or real-time bacterial detection, and offers sensitivity and specificity unchallenged by traditional/conventional methods.

Mycotoxin Detection

The problem of mycotoxin contamination in food including fermented foods is a global concern. Mycotoxin contamination is particularly prevalent in developing countries in tropical areas such as in South Asia and Africa. High-performance liquid chromatography (HPLC) and gas chromatography/mass spectrometry (GC/MS) are two of the most widely used methods for the detection and quantification of mycotoxins in developing countries.

These methods, however, are time-consuming, difficult to use and require laboratory facilities. Immunoassays that are economical in use, sensitive and easy to use would greatly facilitate the detection and quantization of mycotoxins. A number of ELISA kits are now commercially available for the detection of aflatoxins, deoxynivalenol, fumonisins, ochratoxins, and zearalenone.

Detection and Identification of Foods and Food Ingredients

The DNA-based identification code system is reliant on polymorphisms at the nucleotide level for the differentiation of living organisms at the variety and species levels. Currently PCR-based methods are used either for the purpose of detecting single nucleotide polymorphisms (SNPs) giving rise to restriction fragment length polymorphisms (RFLPs) or for detecting small sequence length polymorphisms (SSLPs) often known as Variable Number Tandem Repeats (VNTRs).

These methods facilitate the identification of unique polymorphisms of a variety of food commodities and can be used in the identification of their source or origin. These unique polymorphisms are often referred to as DNA barcodes. The DNA barcode is used for the identification of specific varieties in food detection and in food traceability.

The DNA barcode has been used for the identification of many products for export in countries such as Thailand, China, Brazil, Cuba and Argentina, The DNA barcode of microsatellite markers has also been successfully used in differentiating and identifying fermented products such as premium wines, cheeses and sausages on the basis of their origins. Basmati rice varieties and olive cultivars used in olive oil production have also been differentiated.

ANALYSIS OF THE REASONS FOR SUCCESSES/FAILURES OF APPLICATION OF BIOTECHNOLOGIES IN DEVELOPING COUNTRIES

Socio-economic factors have played a major role in the adoption and application of microbial inoculants in food fermentations. In situations where the cost of food is a major issue, uptake and adoption of improved biotechnologies has been generally slow. Demand for improved inoculants and starter culture development has been triggered by increasing consumer income, education and new market opportunities.

Socio-economics of the Consumer Base

The consumer base of traditionally fermented staple foods in most developing countries is largely poor and disadvantaged. Price, rather than food safety and quality, is therefore a major preoccupation of this group when purchasing food. Fermented foods provide that target group with an affordable source of food, and make a substantial contribution to their food and nutritional security. These foods are generally produced under relatively poor hygienic conditions at the household and village level. Fermentation processing is practised largely as an art in such contexts.

Interventions designed to upgrade processes used in the production of these traditionally fermented staples have been largely carried out through donor-funded projects and have focused primarily on reducing the drudgery associated with the fermentation processes. Improvements have also targeted the upgradation of hygienic conditions of fermentation processes and the introduction of simple and "appropriate" methodologies for the application of inoculants, such as the use of backslopping.

While the uptake of simple backslopping technologies at the household level has, in general, been very good by that target group, the uptake of defined starter cultures has been less successful, owing to cost considerations. Case Study 4.3 on the household level production of Som Fug in Thailand highlights the poor uptake of improved starter culture technologies by household-level processors, primarily on the basis of cost.

With growing incomes and improved levels of education in urban centres across a number of developing countries, dietary habits are changing and a wider variety of foods is being consumed. Fermented foods are no longer the main staples, but are still consumed as side dishes or condiments by that target group.

The demand of that target group for safe food of high quality has begun to re-orient the traditional fermented food sector, and led to improvements in the control of fermentation processes through the development and adoption of defined starter cultures, the implementation of GHPs and HACCP in food fermentation processing, and the development of bioreactor technologies, coupled with appropriate downstream processing to terminate fermentation processes and thus extend the shelf-life of fermented foods.

The packaging of fermented products has also improved. Case Study 4.1 on soy sauce production in Thailand highlights an example of how starter culture development coupled with bioreactor technology has improved yields and the efficiency of fermentation processes, while Case Study 4.2 highlights how consumer demand for safe food led to research and development into starter culture development designed to improve the safety of nham in the marketing chain.

Changing Consumer Demand Trends

Apart from their changing dietary patterns and their demand for safety and quality, higher-income consumers demand convenience and are increasingly concerned about deriving health benefit from the foods they consume. Many of these consumers also show a preference for shopping in supermarkets. Consumer demand for deriving wellness through food consumption has stimulated the development of industrial fermentation processes for the production of functional ingredients such as polyunsaturated fatty acids and pro-biotic cultures for use as food ingredients in developing countries. These functional ingredients are currently applied in the fortification of fermented foods as well as in the production of dietary supplements in countries such as India.

The growth of supermarkets in developing countries has promulgated the need for standardized products of a reasonable shelf-life that meet safety and quality criteria. Packaged fermented products such as kimchi, miso and tempeh, for example, are widely available in supermarkets across Asia. The production of traditional beer in a powdered format and in ready-to-drink containers in Zambia is a very good example of product development that has taken place in response to consumer demand for convenience, both in local and export markets.

Shifting consumer preferences in South Africa, away from basic commodity wine to top-quality wine, is yet another example of how market demand has led to research and biotechnological innovation in the wine industry. Biotechnological innovations in that country are currently focused on the improvement of *Saccharomyces cerevisiae* strains to improve wholesomeness and sensory quality of wines.

The Enabling Environment for Starter Culture Development

A considerable amount of research in developing countries has focused on the identification of starter micro-organisms associated with the fermentation of these staple foods. The greatest strides in starter culture development have, however, been realised in countries that have prioritized the development of technical skills, the infrastructural support base and funding support for research into the upgradation of fermentation processes.

Linkages between research institutions and the manufacturing sector have also been critical to the successful introduction of starter cultures. Case Study 4.1 on soy sauce production exemplifies how success was achieved through such collaboration. Case Study 4.2 on nham production in Thailand also highlights how collaboration between the manufacturing sector and public sector research institutions resulted in the development of improved starter cultures and the uptake of these cultures by nham manufacturers to assure product safety.

Collaborative initiatives among research institutions have also had a major positive impact on biotechnological developments in developing countries. Collaboration among African institutions and their counterparts in the North has greatly facilitated improvements in biotechnological research and capacity development in the area of food biotechnology on the continent.

One major success story in this regard has been collaborative projects involving Burkina Faso, Ghana and institutions in the Netherlands. This programme facilitated the typing and screening of microbial cultures associated with fermented African foods as a basis for starter culture development. Results of this work led to improvements in the production of gari, a fermented cassava product and dawadawa, a fermented legume product. Issues related to the protection of intellectual property rights (IPR) are of growing concern with respect to starter culture development.

Proactive Industrial Strategies

Biotechnology developments have been most successful in areas where proactive approaches are taken by industry. The Thai food industry successfully creates perceived quality by launching new product logos and associating these new products with biotechnology or with the fact that they were developed using traditional biotechnology, such as starter cultures. The goal of the industry is to project an image of itself as producing products of superior quality and safety that represent progressiveness based on a higher level of technology.

Export Opportunities for Fermented Products

Increasing travel due to globalization has changed the eating habits of consumers across the globe. Export markets for fermented foods have grown out of the need to meet the requirements of developing country diaspora in these markets as well as to satisfy growing international demand for niche and ethnic products. Indonesian *tempe* and Oriental soy sauce are well known examples of indigenous fermented foods that have been industrialized and marketed globally. The need to assure the safety and quality of these products in compliance with requirements of importing markets has been a driving force for the upgrading of starter cultures as well as for diagnostic methodologies for verification of their quality and safety. Growing interest and trade in fermented food products is also likely to lead to the greater use of the DNA barcode for identifying the origins of specific fermented food products produced in developing countries.

CASE STUDIES OF APPLICATIONS OF BIOTECHNOLOGIES IN DEVELOPING COUNTRIES

Fermented Soy Sauce Production

This Case Study on the production of soy sauce highlights success in the application of starter culture technology and the use of improved bioreactor technology. It exemplifies the transition of a craft-based production system to a technology-based production system. Research leading to these developments was supported by an international organization, followed by funding support from the Government of Thailand and the Thai soy sauce industry. Developments of the process were largely driven by the demand pull created by a soy sauce industry consortium in Thailand in order to meet market requirements.

Soy sauce production involves a two-step fermentation process that makes use of koji inoculants in the initial phase, followed by moromi inoculants in the second phase. The initial phase of the fermentation involves the soaking of soybeans in water for 1–2 hours, boiling for approximately 17 hours to hydrolyze the protein complex, and the addition of the koji culture *Aspergillus oryzae* for proteolysis of the soy proteins.

Using this traditional method of production, the process of proteolysis takes between 40 hours and seven days depending on the method and the conditions used. The second phase of the fermentation process, which is referred to as a moromi fermentation, involves the addition of brine solution to the koji. *Saccharomyces rouxii,* a salt-tolerant yeast, is the predominant micro-organism in this phase of the fermentation, which lasts as long as 8–12 months.

Moromi fermentations are traditionally conducted in earthenware jars, which often poses a limitation to the manufacturers both in terms of expansion and in terms of production capacity. The soy sauce industry has moved up the ladder of development, from an "art" to a technology-based process through the introduction of defined starter cultures and improvements in the control of the fermentation process. Physical and biological parameters of the fermentation process are controlled through the use of koji and moromi cultures and koji and moromi fermentors.

Use of the koji starter, *Aspergillus flavus* var. *Columnaris,* was found to enhance product safety and uniformity. The introduction of pressure cookers as an innovation for hydrolyzing the soybeans reduced the time required for solubilization from 17 hours to 2.5 hours.

Moreover, the use of starter culture technology facilitated the development of fermentation chambers with controlled temperature and humidity conditions, which resulted in shortening the duration of the fermentation process. The resulting soy sauce had a higher (6 per cent) soluble protein content than that derived from boiled soybeans. These developments resulted in economic gain for the soy sauce industry and greater value added to the product in terms of quality and safety

Traditional Fermented Pork Sausage (Nham)

Nham is an indigenous fermented pork sausage produced in Southeast Asia. It is prepared from ground pork, pork rinds, garlic, cooked rice, salt, chili, sugar, pepper and sodium nitrite.

This Case Study on nham demonstrates how consumer demand for safe food resulted in the commercial use of defined starter cultures, with the collaboration and support of government agencies. The diagnostic role of biotechnology in starter culture development for the tailor making of cultures is also highlighted.

Nham is traditionally consumed as a condiment in the uncooked state in Thailand. It is generally produced using a uncontrolled fermentation process. Fermentation of the product occurs during transportation from the manufacturer to the point of retail. The product is generally retailed under ambient conditions. Traditionally produced nham is considered high risk by Thai health authorities, who require a warning label stating that the product "must be cooked before consumption" on the package.

The first step in the transition to science-based technology for nham fermentations was the development of a starter culture. This starter was subsequently adopted by a nham manufacturer who also implemented HACCP in his operation in order to assure safety and to satisfy the compulsory standard requirements of GMP in the food processing industry imposed by the Thai Food and Drug Administration.

A microbiological hazard profile was developed for nham by the manufacturer in collaboration with scientists from the Ministry of Science, who established that the prevalent pathogens in nham were *Salmonella* spp. (16 per cent), *Staphylococcus aureus* (15 per cent) and *Listeria monocytogenes* (12 per cent). Nitrite, an additive used in nham production was identified as a chemical hazard and the metal clips used for closure of the package were identified as physical hazards. A HACCP plan which included four critical control points was developed for nham.

The critical control point on nitrite was monitored by checking the pre-weighed nitrite prior to adding it to the product formulation. Scientific data generated through the conduct of studies on starter cultures showed that a rapid increase in acidity within 36–48 hours of fermentation inhibited the growth of bacterial pathogens such as *Staphylococcus aureus* and *Salmonella* spp. With the application of these starter cultures, the final product was sent to retailers after the fermentation reached its end-point (pH< 4. 5). An innovative pH indicator which undergoes a colour change on attainment of the end-point of the fermentation process (pH<4. 5) was included in the package. With these innovations and the implementation of a HACCP plan, local health authorities waived the requirement for the warning "must cook before consumption" on the package.

This authorization was seen by the public as an endorsement of product quality and safety by the health authority. Subsequent to these developments, three medium-sized manufacturers followed suit in adopting the improved technology. Recognition of the starter culture technology as a food safety measure by the health authority was, of itself, an effective public awareness campaign.

RAPD markers were used for the molecular typing of approximately 100 bacterial strains at 12-hour intervals during nham fermentations. These studies resulted in the development and commercialization of three different starter formulae for use by larger manufacturers of nham. These starter cultures are marketed in a liquid form which requires refrigeration.

Dried starter cultures have a shelf life of one month at ambient temperature. Further innovations have led to the incorporation of local yeast extracts into starter culture development, resulting in a 20-to 30-fold reduction in cost. The adoption of starter culture technology in nham fermentations has had a positive impact on the industry in terms of safety assurance to consumers and product consistency.

Traditional Fermented Fish Paste–Som Fug

Som Fug is a traditional fermented minced fish cake. It is considered a healthy and highly nutritious product, and is an excellent source of protein (protein content: 15.7 per cent, fat: 3.2 per cent and total carbohydrates: 4 per cent). It is produced using a spontaneous microbial fermentation process similar to that used for producing nham and many other Southeast Asian fermented foods. This Case Study demonstrates that the uptake and use of starter culture technologies is still largely contingent on cost considerations and consumer appreciation of the nutritional value of the product.

Compositionally, Som Fug consists of minced freshwater fish (mud carp, *Cirrhina microlepis*) 84 per cent (w/w), garlic 8 per cent, water 4 per cent, salt 2 per cent, boiled rice 1 per cent, sucrose 0.1 per cent and black pepper. It is fermented for about 2–4 days at ambient temperature. Lactic acid bacteria are the dominant microflora associated with the fermentation. RAPD-PCR analyses determined that the garlic fermenting lactic acid bacteria associated with Som Fug fermentations belonged to *Lactobacillus pentosus* and *Lact. plantarum*. Furthermore, the studies also concluded that fructans from garlic are important carbon sources which catalyze the fermentation of Som Fug. The studies of Som Fug illustrate the high discriminatory power of biotechnology in differentiating lactic acid bacteria at the strain level. The Som Fug industry did not see the benefit of implementing starter-culture technology.

Although the important micro-organisms for Som Fug fermentation had been identified, there were no attempts to develop starter cultures. One major reason for the lack of development of starter-culture technology was the widespread production of Som Fug at the household level. Household manufacturers do not see the benefit of starter-culture technology but, rather, view starter cultures as a burden to the cost of production. Moreover, there is no scientific information to substantiate the nutritional value of Som Fug and hence there is very little public awareness of the nutritional value of the product.

Flavour Production from Alkaline-fermented Beans

This Case Study on the indigenous fermentation of the locust bean, dawadawa (fermented locust bean), is a classic example of how traditional fermentations can be exploited for the production of high-value products such as flavour compounds. The work, however, was undertaken by a large cooperation with little involvement of local researchers. Returns on commercial successes derived from this study did not go back to the people who invented the traditional method of producing this indigenous fermented food. This Case Study, therefore, serves to highlight the critical issue of IPR of traditional production systems.

Dawadawa is produced by alkaline fermentation of the African fermented locust bean. It is an important condiment in the West/Central African Savannah region. Similar fermented food products can be found throughout Africa, with regional differences in the raw materials used as processing inputs or in postprocessing operations.

Similarly fermented products are referred to as "kinda" in Sierra Leone, "iru" in coastal Nigeria, "soumbara" in Gambia and Burkina Faso, and "kpalugu" in parts of Ghana. Foods produced by alkaline fermentation in other parts of the world include "natto" in Japan, "*thua noa*" in Thailand and "kinema" in India. These are mainly used as culinary products to enhance or intensify meatiness in soups, sauces and other prepared dishes.

The production of dawadawa involves extensive boiling and dehulling of the beans, followed by further boiling to facilitate softening. Spontaneous fermentation of the softened beans is subsequently allowed to take place over 2–4 days. Micro-organisms associated with the fermentation include *Bacillus subtilis, B. pumilus, B. licheniformis* and *Staphylococcus saprophyticus*. During the fermentation process, the pH increases from near neutral to approximately

8.0, temperature increases from 25 °C to 45°C and moisture increases from 43 per cent to 56 per cent. At the same time, a five-fold increase in free amino acids takes place, and glutamate, a flavour enhancer, increases five-fold during the process. Mechanisms of flavour production during the fermentation process, as well as flavour principles generated during dawadawa fermentation processing, have been studied by international food manufacturers and been used as a basis for the development of flavours for incorporation in bouillon-type products.

LOOKING FORWARD: PREPARING FOR THE FUTURE

A KEY ISSUE IN THE SECTOR WHERE THE APPLICATION OF BIOTECHNOLOGIES COULD BE USEFUL

Emerging Pathogens

The identification of infectious agents requires high-end technologies which are not usually available in developing countries. Developing countries must, therefore, seek assistance from countries with higher calibre technologies in order to characterize the infectious agents, put in place surveillance and monitoring systems and develop strategies to contain the disease(s). Biotechnology can play a key role in facilitating the characterization of new emerging pathogens. Traditional cultural methods for the detection and enumeration of microbial pathogens are tedious and require at least 12–18 hours for the realization of results. By that time, the food products would have been distributed to retailers or consumers. Immunoassay diagnostic kits facilitate near-real-time monitoring, sensitivity, versatility and ease of use. The emergence of multi-antibiotic resistance traits is prevalent in intensive farming in developing countries due to the abuse of antibiotics. The spread of multi-antibiotic resistant micro-organisms poses public health concerns, because pathogens exhibiting such resistance would be difficult to control with the use of currently available antibiotics. The rapid detection of these pathogens, with high sensitivity, is one way of monitoring and containing the spread of multi-antibiotic resistant traits. A strategic approach being employed by some is the development of affinity biosensors with an antibiotic resistant nucleotide sequence as the detection probe.

IDENTIFYING OPTIONS FOR DEVELOPING COUNTRIES

It is important that countries recognize the potential of fermented foods and prioritize actions to assure their safety, quality and availability. Based on the stocktaking exercise in this document, a number of specific options can be identified for developing countries to help them make informed decisions regarding adoption of biotechnologies in food processing and in food safety for the future.

Regulatory and Policy Issues

- Governments must be committed to protecting consumer health and interests, and to ensuring fair practices in the food sector.
- There has to be consensus at the highest levels of government on the importance of food safety, and the provision of adequate resources for this purpose.
- Government policy that is based on an integrated food-chain approach is science-based, transparent and includes the participation of all the stakeholders from farm to table must be put in place.
- The importance of the regional and international dimensions of the use of biotechnologies in food processing and safety must be recognized.
- Priority must be accorded to promoting fermented foods in the food-security agendas of countries.
- Governments must also provide an enabling environment that is supportive of the growth and development of upstream fermentation processes such as the production of high-value fermented products, such as enzymes, functional-food ingredients and food additives.

International Cooperation and Harmonization

- The organization and implementation of regional and international fora are critical requirements for the enhancement of national organizational capability and performance and for the facilitation of international co-operation. Further, the setting up of administrative structures with clearly defined roles, responsibilities and accountabilities could efficiently govern processed foods and safety issues.
- Biotechnology-based Standard Operating Procedures (SOPs) for food safety should also be documented for use in authorized laboratories.
- National food control databases for the systematic collection, reporting and analysis of food-related data (food inspection, analysis, etc.) with set regulations and standards based on sound science and in accordance with international recommendations (Codex) are key requirements.

Education Policy

- While the consumption of fermented foods is growing in popularity among higher-income consumers thanks to increasing interest in wellness through diet, the consumption of fermented foods by lower-income consumers in many developing countries is perceived to be a backward practice.
  - Strategies should therefore be developed for the dissemination of knowledge about food biotechnology and, particularly,

fermented foods. Targeted consumer education on the benefits of consuming fermented food products and on applying good practice in their production is required.

- Food biotechnology should be included in educational curricula in order to improve the knowledge base in countries on the contribution of fermented foods to food and nutritional security and to generate awareness of the growing market opportunities for fermented foods and high-value products derived from fermentation processes.

Information-sharing

Access to specialized technical information on biotechnology and biotechnological developments in the food processing sector are critical and necessary inputs and support systems for guiding and orienting the research agendas of countries. The necessary information systems should therefore be developed to facilitate rapid access to information on biotechnological developments across both the developed and the developing world.

Legislation and Policy on Technologies

Expertise in legislation and technology licensing, as well as knowledge about how to nurture innovation and turn it into business ventures, are critical requirements for developing countries. Successful technology transfer requires all of these elements and an environment that is conducive to innovation. Government policy in developing countries should therefore prioritize technology transfer that helps create new business ventures, an approach that requires government support such as tax incentives and infrastructure investment.

Intellectual Property Rights (IPR)

- Many of the traditional fermentation processes applied in developing countries are based on traditional knowledge. Enhanced technical and scientific information is required in order to claim ownership of the traditional knowledge of the craft of indigenous fermented foods. Lack of technical knowledge has resulted in the failure to realise the benefits of the industrialization of indigenous fermented foods by individuals who are the rightful owners of the technology.
- Greater focus is required on issues of relevance to IPR and on the characterization of microbial strains involved in traditional fermentation processes. Emphasis must be placed on IPR education for scientists. National governments should put in place the requisite infrastructure for IPR to facilitate the process. At the institutional level, this infrastructure would include technology management

offices for assisting scientists in procedures relating to intellectual property matters. The processes used in the more advanced areas of agricultural biotechnology are generally covered by IPR, and the rights are generally owned by parties in developed countries.

Communication and Consumer Perceptions

- Communication between various stakeholders is critical in proactively engaging with consumers. Communication must be established with the public at large on processed food and associated hazards. Communication gradually builds confidence and will be critical to advancing the application of biotechnologies in food processing and safety. The primary role of communication in this respect is to ensure that information and opinions from all stakeholders are incorporated in the discussion and decision-making process. The need for specific standards or related texts and the procedures followed to determine them should also be clearly outlined. The process, therefore, should be transparent.
  - Public awareness and education are critical to the success of food bioprocessing and food safety in developing countries.
  - Greater attention must be directed towards understanding consumer and producer (processor) perceptions on food safety and quality in developing countries.
  - If foods are to be promoted as being safe and healthy, their nutritional and safety attributes must be transparently demonstrated by presenting scientific data to substantiate the nutritional and health benefits and by applying good manufacturing/hygiene practice and HACCP as safety measures to ensure that issues of consumer concern are addressed.

Technical Capacities and Technology Transfer

- Traditional fermented foods should be viewed as valuable assets. Governments should capitalize on these assets and add value to them by supporting research, education and development, while building on and developing the indigenous knowledge base on food fermentations.
  - Government agencies in developing countries should focus on the development of technical capacities to deal with emerging technical issues.
  - The technical capacities of academic and research institutes should be strengthened in the fields of food biotechnology, food processing, bioprocess engineering and food safety through training and exchange programmes for researchers. Such programmes should emphasize collaboration with both

developed and developing country institutions engaged in work on food biotechnology, starter culture development, bioprocess engineering and food safety.

- Training capabilities in food biotechnology and food safety should be developed within developing country institutions through the introduction of degree courses in order to broaden the in-country technical support base for food bioprocess development. Given the similarities among fermentation processes across regions, an inventory of institutions engaged in food biotechnology in developing countries would be an asset in facilitating networking among institutions. Food processors, policy-makers and equipment manufacturers should also be integrated into the networking activities.
- The development of appropriate levels of bioreactor technology with control bioprocess parameters will be necessary to improve the hygienic conditions of the fermentation processes.
- Research and infrastructural development to enable the cost-effective production of defined starter cultures in a stable format (*i.e.* cultures which do not require refrigeration and have prolonged shelf-life under ambient conditions) should be prioritized.
- Infrastructure development to facilitate the transfer and adaptation of fermentation technologies developed at the laboratory level to the household and village and, where necessary, the enterprise level should be prioritized.
- Appropriate levels of equipment will also be required to facilitate the downstream processing of these products.
- Traceability systems that facilitate the differentiation and identification of food products should be prioritized in order to broaden market opportunities for these products.
- A food-chain approach to assuring food safety should be prioritized by governments.

- Food safety management systems should be strengthened by implementing systematic food safety measures such as GHP, GMP and HACCP in food fermentation operations. Diagnostic kits are important tools for monitoring and verifying the level of sanitation in processing plants.
  - Highly sensitive and rapid diagnostic kits are invaluable for monitoring and rapidly detecting chemical and microbiological hazards that pose a threat to human health, with high precision and sensitivity. The development of low-cost diagnostic kits suitable for use by small processors would greatly facilitate food-safety monitoring. Development should target the realization of multiplex

diagnostic systems with the capacity to detect several pathogens or many chemical contaminants using a single diagnostic kit. The development of diagnostic kits at a national level could further reduce their cost of production. Given the regional specificity of bacterial pathogens at the species and subspecies levels, such diagnostic kits should be developed with specificity and sensitivity to the species or subspecies that are prevalent in a specific region. Investment is therefore needed for the development of expertise, facilities and the infrastructure for the mass production of antibodies, cell culture technology and for the formation of technical know-how on assembling the requisite components of diagnostic kits.

– The development of national hazard-profile databases that document the prevalent pathogens in different regions will be critical. Such information would be useful for further research into the development of diagnostic kits with high precision and sensitivity and in implementing HACCP as well as risk assessment research. The culture collection of identified infectious agents in the hazard profiles could play an important role for specific antibody production for use in the development of immunoassay diagnostic kits.

IDENTIFYING PRIORITIES FOR ACTION FOR THE INTERNATIONAL COMMUNITY

The last decade has witnessed considerable change with respect to the application of biotechnology in food processing and food-safety applications. Market forces have been the major drivers of change in the food sector of developing countries. Modern biotechnological tools are likely to play a greater role in the development of efficient science-based processes for food processing and safety in order to respond to consumer demand. The production of high-value fermented products such as enzymes, functional food ingredients and food additives is likely to continue to increase in developing countries. The international community (FAO, UN organizations, NGOs, donors and development agencies) can play a major role in assisting developing countries to maximize the benefit to be derived from food bioprocessing. The adoption of biotechnology-based methods in food processing and for food safety and quality monitoring is dependent on several factors that include capacity-building in technical and regulatory areas, policy formulation, regulatory frameworks and regional networks. On the basis of analysis in this document, a number of priority areas to be supported by the international community are presented below.

Capacity-building and Human Resource Development

- Support for basic and advanced education.

- Prioritization of specific areas for investment.
- Development of policies, priorities and action programmes that promote food fermentation as a means of addressing food security.
- Support for human resource development in a range of scientific disciplines–food biotechnology, food safety, bioengineering and enzyme technology.
- Support for capacity-building initiatives for household-level, small- and edium-scale processors of fermented foods.
- Support for IPR development.

Technology Transfer and Support for Research and Development

- Improvement of the relevance of national research to the needs of the food sector in developing countries.
- Enhancement of competitiveness and the creation of an enabling environment that is conducive to private-sector investment in research, development and innovation for the upgrading of food fermentation processes to respond to market demand.
- Establishment and strengthening of the research and infrastructural support base for work on starter culture development, bioreactor design and for the development of diagnostic equipment for monitoring food safety and traceability. This infrastructural support base would include laboratories, laboratory equipment, cell bank facilities for the proper preservation and storage of microbial culture preparations.
- Development of scientific data to substantiate the nutritional, health and health-benefit claims associated with fermented foods.
- Establishment of pilot processing facilities for the scale-up and testing of technologies developed in order to facilitate their adoption.

Networking and Clusters

- Support for the development of regulatory frameworks for food safety.
- Support for North-South and South-South training and exchange on food biotechnologies, bioprocess engineering and food safety.
- Promotion and facilitation of networking among scientists, researchers, small-and medium-scale food processors and the retail sector in order to facilitate knowledge and information-sharing.
- Support for leveraging the traditional knowledge base in the upgrading of food-fermentation processing operations.

4

Dairy Waste

INTRODUCTION

The rapid growth in the size of dairy operations has resulted in new laws and regulations governing the handling and disposal of manure. Requirements for nutrient management plans, manure solids disposal, and odour control make it necessary that new manure management approaches be considered. One of the more promising methods is anaerobic digestion. Anaerobic digestion is a natural process that converts biomass to energy.

Biomass is any organic material that comes from plants, animals or their wastes. Anaerobic digestion has been used for over 100 years to stabilize municipal sewage and a wide variety of industrial wastes. Most municipal wastewater treatment plants use anaerobic digestion to convert waste solids to gas.

The anaerobic process removes a vast majority of the odorous compounds. It also significantly reduces the pathogens present in the slurry. Over the past 25 years, anaerobic digestion processes have been developed and applied to a wide array of industrial and agricultural wastes.

It is the preferred waste treatment process since it produces, rather than consumes, energy and can be carried out in relatively small, enclosed tanks. The products of anaerobic digestion have value and can be sold to offset treatment costs. Anaerobic digestion provides a variety of benefits.

The environmental benefits include:

- Odours are significantly reduced or eliminated.
- Flies are substantially reduced.
- A relatively clean liquid for flushing and irrigation can be produced.
- Pathogens are substantially reduced in the liquid and solid products.
- Greenhouse gas emissions are reduced.
- And finally, nonpoint source pollution is substantially reduced.

On the economic side, additional benefits are provided:

- The time devoted to moving, handling, and processing manure is minimized.

- Biogas is produced for heat or electrical power.
- Waste heat can be used to meet the heating and cooling requirements of the dairy.
- Concentrating nutrients to a relatively small volume for export from the site can reduce the land required for liquid waste application.
- The rich fertilizer can be produced for sale to the public, nurseries, or other crop producers.
- Income can be obtained from the processing of imported wastes, the sale of organic nutrients, greenhouse gas credits, and the sale of power.
- Power tax credits may be available for each kWh of power produced.
- Greenhouse tax credits may become available for each ton of carbon recycled.
- Finally the power generated is "distributed power" which minimizes the need to modify the power grid. The impact of new power on the power grid is minimized.

In order to achieve the benefits of anaerobic digestion, the treatment facility must be integrated into the dairy operation. Unfortunately, no single dairy can serve as a model for a manure treatment facility. The operation of the dairy will establish the digester loading and the energy generated from the system.

The anaerobic facility must be designed to meet the individual characteristics of each dairy. This manual provides an introduction to the anaerobic digestion of dairy manure. It is divided into three parts.

The first describes the operation and waste management practices of Idaho dairies. The second introduces anaerobic digestion and the anaerobic digestion processes suitable for dairy waste. The third presents typical design applications for different types of dairies and establishes the cost and benefits of the facilities.

DAIRY OPERATIONS

Dairy operations significantly affect the quantity and quality of manure that may be delivered to the anaerobic digestion system. In addition to the number of milk and dry cows, the housing, transport, manure separation, and bedding systems used by the dairy establishes the amount of material that must be handled and the amount of energy produced.

HOUSING SYSTEM

Confined dairy animals may be housed in a variety of systems. Commonly used housing systems include free stalls, corrals with paved feed lanes, and open lot systems. Milk cows, dry cows and heifers may be housed in free stalls, corrals, and open-lots on the same dairy. The type of housing used determines the quantity of manure that can be economically collected.

Free Stall Barns

Free stalls are currently the most popular method for housing large dairy herds. Free stall housing provides a means for collecting essentially all of the manure.

Fig. Typical Free Stall Barn with Center Feed Lane

Corrals

Corral systems with paved feed lanes are also commonly used. The manure deposited in the feed lanes can be scraped or flushed daily. From 40 to 55-per cent of the excreted manure may be deposited and collected from the corral feed lane. The balance of the manure may be deposited in the milk barn (10 to 15 per cent) or the open lot (30 to 50 per cent).

Fig. Corral with Paved Feed Lane for Scrape or Vacuum Collection

Typically the manure deposited in the open lot is removed two to three times a year. It may have little net energy value after being stored in the open lot over prolonged periods of time. For corral systems one must make a reasonable determination of the recoverable manure deposited in the feed lane, corral, and milk barn. Corral systems also use a considerable amount of bedding material during the winter months. The straw bedding is generally removed in the spring and placed on the fields prior to spring planting.

Milk Barn

Dairy cows are milked two to three times a day. The cows are moved from their stalls to the milk parlor holding area. The milk parlor and holding area are normally flushed with fresh water. From 10 to 15 per cent of the manure is deposited in the milk parlor. In addition to the manure that is flushed, the cows may be washed with a sprinkler system. Warm water that is produced by the refrigeration compressors, vacuum pumps, and milk cooling system may be used for drinking water, manure flushing or washing the cows. It has been estimated that 5 to 150 gallons of fresh water per milk cow is used in the milking center. More common values are 10 to 30 gallons of fresh water per milk cow. The quantity and quality of water discharge from the milk parlor must be accurately measured. In many cases, the waste deposited in the milk barn is processed in a separate waste management system.

Open Lot

Fig. Open Lot System

In open lot systems the manure is deposited on the ground and scraped into piles. The manure is removed infrequently (once or twice a year). A significant amount of manure degradation occurs resulting in greenhouse gas emissions. In many cases, the open lot degradation produces manure that has little or no net energy value.

TRANSPORT SYSTEM

The commonly used manure transport systems are flush, scrape, vacuum, and loader systems. In free stall barns the manure can be flushed, scraped, or vacuum collected.

Flush Systems

If a flush system is used the manure is substantially diluted. The quantity of water used in a flush system depends on the width, length, and slope of the flush isle. The feed isles are generally 14 feet wide while the back isles are generally 10 feet wide. The slope varies between one and two per cent. A flush system will generally reduce the concentration of manure from 12 1/2 per cent solids, "as excreted", to less than one per cent solids in the flush water. Flush systems are however more economical and less labour-intensive than scrape or vacuum systems.

Fig. Free Stall Flush System-Flushing Feed Lane

Scrape Systems

Scrape systems are simply systems that collect the manure by scraping it to a sump. Under normal weather conditions the scraped manure has approximately the same consistency as the "as excreted" manure. During the warm dry summer manure may be dewatered on the slab.

Front-end Loader

Front-end loaders are used to stack and remove corral bedding and manure.

Vacuum Systems

Vacuum systems collect "as excreted" manure with a vacuum truck. Generally, the trucks collect approximately 4000 gallons per load. The manure can be hauled to a disposal site rather than to an intermediate sump. Vacuum collection is a slow and tedious process. The advantage is that the collected manure is undiluted and approximately equal to the "as excreted" concentration.

Fig. Vacuum Truck Collecting Manure Solids

BEDDING

The type of bedding used can significantly alter the characteristics of the manure being treated. Typically straw, wood chips, sand, or compost are used as bedding material. In some cases paper mixed with sawdust is used. Compost usually has some sand mixed with the organic constituents. If composting is carried out on dirt lots, a significant amount of sand and silt may be incorporated into the compost.

Since anaerobic digestion will not degrade the wood chips, sand, or silt, it is necessary to remove those constituents prior to, or during anaerobic digestion process. The quantity of non-degradable, organic and inorganic material can significantly impact the performance of the anaerobic digester. The quantity of bedding added to the manure is a function of the design and operation of the dairy. Generally only the "kick-out" from the stalls is added to the manure. The quantity that is "kicked-out" is a function of the design of the dairy housing system as well as the type of bedding used.

MANURE PROCESSING

Each dairy has its own manure processing system. Scraped or flushed manure may be processed in a system separate from the milk barn waste, or the collected manure waste may be processed with the milk barn waste. In general, current manure processing consists of macerating the waste with a chopper pump, screening the waste to remove the organic fibres, followed by sedimentation to remove the sand, silt, and organic settable particles. Much of the degradable manure is removed during the separation processes. Up to 80 per cent of the COD and 30 per cent of the total Nitrogen and Phosphorous can be lost in the solids removed by the screen and sedimentation process. Detailed sampling and analysis is required to confirm losses.

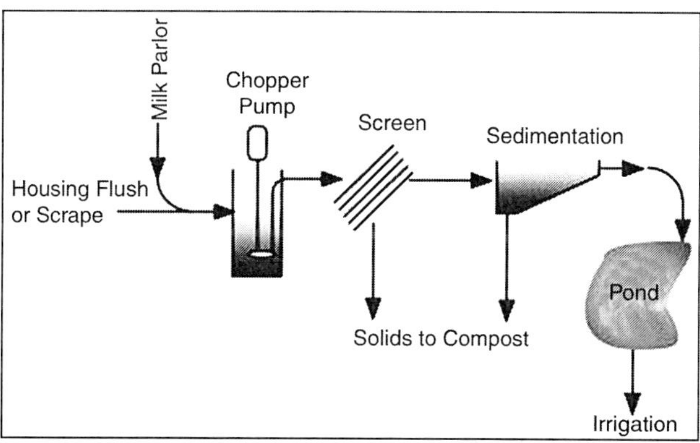

Fig. Conventional Manure Handling

Holding Tanks and Chopper Pumps

A wide variety of holding tanks and chopper pumps are used throughout the dairy industry. Typically, the tanks are relatively small but in some cases they are designed to hold several hours of flush water.

Fig. Typical Manure Sump with Chopper Pump

Primary Screens

An equally wide variety of screening systems are used. In many cases the screens are housed in separate enclosures to prevent freezing during the winter months. Outdoor screens are generally problematic during cold weather months. Fan separators are also used to provide efficient separation of the fibrous solids.

The primary screens will remove a significant amount of degradable organic material that could be converted to gas in an anaerobic digester. The screened materials are generally used to produce bedding after being composted for the required time periods.

Fig. Primary Screens Background with Gravity Separators Foreground

Gravity Separators

Gravity separators varying in size from 10 feet wide by 30′ long to 24 feet wide by 80 feet long are usually placed after the primary screens. The purpose of the gravity separator is to remove the sands and silt present in the waste stream. If gravity separators are used without screening a thick mat of straw and fibres may develop on top of the gravity separator.

Fig. Floating Solids on Top of Gravity Separator

The gravity separators often incorporate weeping walls for the removal of liquid from the sedimentation chamber. The ability of the weeping wall to remove liquid waste depends on the periodic cleaning of the perforations to maintain flow. In many cases the gravity separators remove a significant amount of degradable organic material that could be utilized to produce gas. The COD test is a direct measure of the quantity of material that could be converted to methane gas.

Recent tests have established that screen and gravity separators can remove 75 per cent to 80 per cent of the COD present in the waste stream. In one test the dairy parlor COD was reduced from 31,000 mg/l to 8,600 mg/l in the effluent from the gravity separator. In another the flush water influent to a separator system was 10,900 mg/L while the effluent was 1,800 mg/L. While a significant portion of the organic carbon (COD) is retained with the separated solids, an equal percentage of the nitrogen and phosphorus is not. The separation process alters the carbon to nitrogen ratio of both the liquid and solids streams.

Fig. Weeping Wall with Clogged Holes

The sedimentation process concentrates the organic solids, which are periodically removed. A recent analysis showed the flush water had a COD concentration of 25, 500 mg/l while the concentrated solids from the separator had a COD of 115, 800 mg/l.

Fig. Organic Solids on Top of Gravity Separator

Anaerobic decomposition of settled solids can be observed in separators that have a surface covered with methane gas bubbles. It is clear that existing solids handling practices contribute to greenhouse gas emissions and prevent efficient energy recovery from manure waste.

Fig. Gravity Separator with Anaerobic Decomposition of Organic Solid

Primary Holding Ponds

Most dairies will discharge the screened and settled waste to a primary holding pond.

Fig. Partially Empty Primary Holding Pond Showing Sediment

The primary holding pond is a secondary sedimentation basin where the fine solids are separated from the liquid waste. Eventually the fine solids must be removed from the bottom of the primary holding pond. Odours generally accompany the removal of solids.

Secondary Holding Ponds

It is common to have a number of holding ponds that provide the required detention time (180 days) following the primary holding pond. The irrigation and flush pumps are normally installed in one or more of these ponds.

Fig. Typical Secondary Holding Pond

REVIEW

Two separate waste streams, the milk parlor and confinement area wastes, makeup the dairy waste that can be treated through anaerobic digestion. The type of bedding used, as well as the manure transport, and subsequent manure processing will change the characteristics of both waste streams. The dilution of waste will require larger anaerobic digestion facilities. The removal of organics through screening and sedimentation will reduce the quantity of organic solids that can be converted to gas in the digester. The presence of sand and silts will clog pipes, damage equipment, and fill anaerobic digestion tanks.

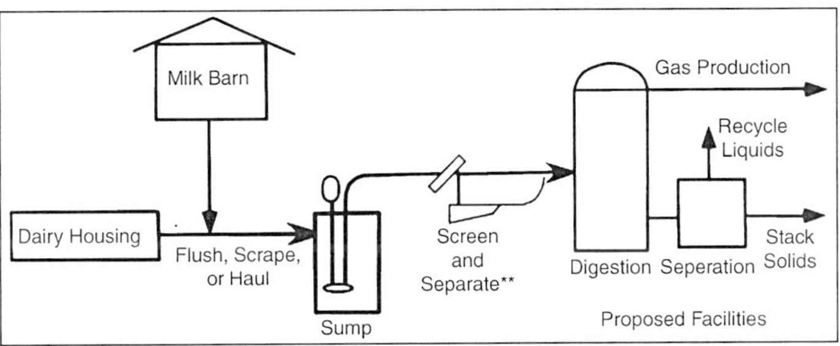

Fig. Integration of Anaerobic Digestion in Dairy Waste Stream

Notes:

** Screen and Separate may be By-passed

Sand can only be removed from dilute waste streams. Thick slurries retain sand that precipitate in the digester when the organics are converted to gas and the solids concentration is reduced. If thick slurries are processed in an anaerobic digester, intense mixing is required to maintain the solids in suspension. Modification of existing dairy management practices may be required to achieve the full benefits of anaerobic digestion.

Figure shows how a solid waste management facility can be incorporated in an existing dairy waste-processing stream. If low or moderate concentrations of sand are present the entire waste stream may be discharged to an anaerobic digester, bypassing the existing screen and gravity separators. If high concentrations of sand are present, the existing gravity separators may remain in place. Under such conditions, a reduced quantity of organics will be converted to gas.

ANAEROBIC DIGESTION

Anaerobic digestion is the breakdown of organic material by a microbial population that lives in an oxygen free environment. Anaerobic means literally "without air". When organic matter is decomposed in an anaerobic environment the bacteria produce a mixture of methane and carbon dioxide gas. Anaerobic digestion treats waste by converting putrid organic materials to carbon dioxide and methane gas. This gas is referred to as biogas. The biogas can be used to produce both electrical power and heat.

The conversion of solids to biogas results in a much smaller quantity of solids that must be disposed. During the anaerobic treatment process, organic nitrogen compounds are converted to ammonia, sulfur compounds are converted to hydrogen sulfide, phosphorus to orthophosphates, and calcium, magnesium, and sodium are converted to a variety of salts. Through proper operation, the inorganic constituents can be converted to a variety of beneficial products.

The end products of anaerobic digestion are natural gas (methane) for energy production, heat produced from energy production, a nutrient rich organic slurry, and other marketable inorganic products. The effluent containing particulate and soluble organic and inorganic materials can be separated into its particulate and soluble constituents. The particulate solids can be sold or exported from the dairy while the nutrient rich liquids are applied to the land.

BACTERIAL CONSORTIA

Anaerobic digestion is carried out by a group, or consortia of bacteria, working together to convert organic matter to gas and inorganic constituents. The first step of anaerobic digestion is the breakdown of particulate matter to soluble organic constituents that can be processed through the bacterial cell wall. Hydrolysis, or the liquification of insoluble materials is the rate-limiting step in anaerobic digestion of waste slurries.

This step is carried out by a variety of bacteria through the release of extra-cellular enzymes that reside in close proximity to the bacteria. The soluble organic materials that are produced through hydrolysis consist of sugars, fatty acids, and amino acids. Those soluble constituents are converted to carbon dioxide and a variety of short chain organic acids by acid forming bacteria. Other groups of bacteria reduce the hydrogen toxicity by scavenging hydrogen to produce ammonia, hydrogen sulfide, and methane. A group of methanogens converts acetic acid to methane gas. A wide variety of physical, chemical, and biological reactions take place. The bacterial consortia catalyze these reactions. Consequently, the most important factor in converting waste to gas is the bacterial consortia. The bacterial consortia are essentially the "bio-enzymes" that accomplish the desired treatment. A poorly developed or stressed bacterial consortium will not provide the desired conversion of waste to gas and other beneficial products.

FACTORS CONTROLLING THE CONVERSION OF WASTE TO GAS

The rate and efficiency of the anaerobic digestion process is controlled by:

- The type of waste being digested,
- Its concentration,
- Its temperature,
- The presence of toxic materials,
- The pH and alkalinity,
- The hydraulic retention time,
- The solids retention time,
- The ratio of food to microorganisms,
- The rate of digester loading,
- And the rate at which toxic end products of digestion are removed.

Waste Characteristics

All waste constituents are not equally degraded or converted to gas through anaerobic digestion. Anaerobic bacteria do not degrade lignin and some other hydrocarbons. The digestion of waste containing high nitrogen and sulfur concentrations can produce toxic concentrations of ammonia and hydrogen sulfide. Wastes that are not particularly water-soluble will breakdown slowly. Dairy wastes have been reported to degrade slower than swine or poultry manure. The manure production from a typical 1,400-pound milk cow is presented in the table below.

Table. Dairy Manure Production

Manure Produced by a 1,400 pound Cow	
Manure (pounds)	112
Manure (gallons)	13.5
Total Solids (dry pounds)	14
Volatile Solids (dry pounds)	11.9
COD (pounds)	12.5
TKN (pounds)	0.63
Total Phosphorus (pounds)	0.098
Total Potassium (pounds)	0.36

The composition of the manure solids is presented in Table.

Table. Dairy Manure Composition

Component	**% Dry of Matter**
Volatile solids	83.0
Ether Extract	2.6
Cellulose	31.0
Hemicellulose	12.0
Lignin	12.2
Starch	12.5
Crude Protein	12.5
Ammonia	0.5
Acids	0.1

The majority of the volatile solids are composed of cellulose and hemicelluloses. Both are readily converted to methane gas by anaerobic bacteria. As pointed out earlier, lignin will not degrade during anaerobic digestion. Since a substantial portion of the volatile solids in dairy waste is lignin, the percentage of cow manure volatile solids that can be converted to gas is lower when compared to other manure and wastes. The manure characteristics also establish the percentage of carbon dioxide and methane in the biogas produced. Dairy waste biogas will typically be composed of 55 to 65 per cent methane and 35 to 45 per cent carbon dioxide. Trace quantities of hydrogen sulfide and nitrogen will also be present.

Dilution of Waste

The waste characteristics can be altered by simple dilution. Water will reduce the concentration of certain constituents such as nitrogen and sulfur that produce products (ammonia and hydrogen sulfide) that are inhibitory to the anaerobic digestion process. High solids digestion creates high concentrations of end products that inhibit anaerobic decomposition. Therefore, some dilution can have positive effects. The literature indicates that greater reduction efficiencies occur at concentrations of approximately 6 to 7 per cent total solids.

Dairy waste "as excreted" is approximately 12 per cent total solids and 10.5 per cent volatile solids. Most treatment systems operate at a lower solids concentration than the "as excreted" values. Dilution also causes stratification within the digester. Undigested straw forms a thick mat on top of the digester while sand accumulates at the bottom. The optimum waste concentration is based on temperature and the quantity of straw and other constituents that are likely to separate within the anaerobic digester.

It is desirable to keep the separation or stratification in the digester to a minimum. Intense mixing involving the consumption of power may reduce the stratification of dilute waste. The use of flush systems to remove the manure from the dairy barns has major economic advantages to the dairy. Flush systems normally use 100 to 200 gallons per cow, per day of dilution water. The flush volumes required are based on the lane or gutter length, width and slope. The flush water usually contains very low concentrations of total and volatile solids. At 100 gallons per cow of flush water, the waste has only 12.5 per cent of the "as excreted" concentration. At 200 gallons per cow per day of flush water the waste contains only 6.25 per cent of the "as excreted" concentration. Table below presents the waste characteristics using various flush volumes.

Table. Manure Waste Concentration with Various Flush Volumes

"As Excreted (AE) Manure"	**Manure with 100 gal per Cow Flush Water**	**Manure with 200 gal per Cow Flush Water**	
Gallons for 1000 milk cows	14,267	114,000	214,000
Total Solids Concentration (mg/l)	120,000	15,000	8,000
Volatile solids Concentration (mg/l)	102,000	12,750	6,800

COD Concentration (mg//)	129,400	16,176	8,627
TKN Concentration (mg/l)	5,294	662	353
Total P Concentration (mg/l)	824	103	55

The milk parlor also produces a substantial amount of dilute waste. Approximately 15 per cent of the animal manure is deposited in the milk parlor. The resulting milk parlor waste has a composition similar to the flush waste presented in Table.

Foreign Materials

Addition of foreign materials such as animal bedding, sand and silt can have a significant impact on the anaerobic digestion process. For example, the poor performance of the Monroe, WA dairy digester was attributed to the use of cedar wood chip bedding. The quantity and quality of the bedding material added to the manure will have a significant impact on the anaerobic digestion of dairy waste. Sand and silt must be removed before anaerobic digestion. If it is not removed before digestion it must be suspended during the digestion process.

Toxic Materials

Toxic materials such as fungicides and antibacterial agents can have an adverse effect on anaerobic digestion. The anaerobic process can handle small quantities of toxic materials without difficulty. Storage containers for fungicides and antibacterial agents should be placed at locations that will not discharge to the anaerobic digester.

Nutrients

Bacteria require a sufficient concentration of nutrients to achieve optimum growth. The carbon to nitrogen ratio in the waste should be less than 43. The carbon to phosphorus ratio should be less than 187. Hills and Roberts showed that a non-lignin C/N ratio of 20 to 25 is optimum for digester performance. Typically, "as excreted manure has a C/N ratio of 10.

Temperature

The anaerobic bacterial consortia function under three temperature ranges. Psychrophilic temperatures of less than 68 degrees Fahrenheit produce the least amount of bacterial action. Mesophilic digestion occurs between 68 degrees and 105 degrees Fahrenheit. Thermophilic digestion occurs between 110 degrees Fahrenheit and 160 degrees Fahrenheit. The optimum mesophilic temperature is between 95 and 98 degrees Fahrenheit. The optimum thermophilic temperature is between 140 and 145 degrees Fahrenheit. The rate of bacterial growth and waste degradation is faster under thermophilic conditions. On the other hand, thermophilic digestion produces an odorous effluent when compared to mesophilic digestion.

Thermophilic digestion substantially increases the heat energy required for the process. In most cases, sufficient heat is not available to operate in the thermophilic range. This is especially true if flush systems are used or the milk parlor waste is mixed with the scraped manure. Large quantities of dilution flush water must be heated to the digester's operating temperature. During cold weather, control of the flush volume is critical in maintaining adequate digester temperatures.

Seasonal and diurnal temperature fluctuations significantly affect anaerobic digestion and the quantities of gas produced. Bacterial storage and operational controls must be incorporated in the process design to maintain process stability under a variety of temperature conditions. Temperature is a universal process variable. It influences the rate of bacterial action as well as the quantity of moisture in the biogas. The biogas moisture content increases exponentially with temperature. Temperature also influences the quantity of gas and volatile organic substances dissolved in solution as well as the concentration of ammonia and hydrogen sulfide gas.

pH

Methane producing bacteria require a neutral to slightly alkaline environment in order to produce methane. Acid forming bacteria grow much faster than methane forming bacteria. If acid-producing bacteria grow too fast, they may produce more acid than the methane forming bacteria can consume. Excess acid builds up in the system. The pH drops, and the system may become unbalanced, inhibiting the activity of methane forming bacteria. Methane production may stop entirely. Maintenance of a large active quantity of methane producing bacteria prevents pH instability. Retained biomass systems are inherently more stable than bacterial growth based systems such as completely mixed and plug flow digesters.

Hydraulic Retention Time (HRT)

Most anaerobic systems are designed to retain the waste for a fixed number of days. The number of days the materials stays in the tank is called the Hydraulic Retention Time or HRT. The Hydraulic Retention Time equals the volume of the tank divided by the daily flow (HRT=V/Q). The hydraulic retention time is important since it establishes the quantity of time available for bacterial growth and subsequent conversion of the organic material to gas. A direct relationship exists between the hydraulic retention time and the volatile solids converted to gas.

Solids Retention Time (SRT)

The Solids Retention Time (SRT) is the most important factor controlling the conversion of solids to gas. It is also the most important factor in maintaining digester stability. Although the calculation of the solids retention time is often improperly stated, it is the quantity of solids maintained in the digester divided by the quantity of solids wasted each day.

Where V is the digester volume; Cd is the solids concentration in the digester; Qw is the volume wasted each day and Cw is the solids concentration of the waste. In a conventional completely mixed, or plug flow digester, the HRT equals the SRT. However, in a variety of retained biomass reactors the SRT exceeds the HRT. As a result, the retained biomass digesters can be much smaller while achieving the same solids conversion to gas.

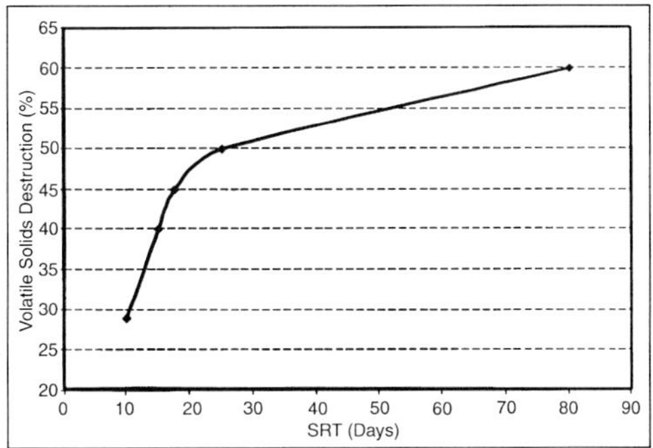

Fig. Dairy Waste Volatile Solids Destruction

The volatile solids conversion to gas is a function of SRT (Solids Retention Time) rather than HRT. At a low SRT sufficient time is not available for the bacteria to grow and replace the bacteria lost in the effluent. If the rate of bacterial loss exceeds the rate of bacteria growth, "wash-out" occurs. The SRT at which "wash-out" begins to occur is the "critical SRT".

Jewel established that a maximum of 65 per cent of dairy manure's volatile solids could be converted to gas with long solids retention times. Burke established that 65 to 67 per cent of dairy manure COD could be converted to gas.

Long retention times are required for the conversion of cellulose to gas. The goal of process engineers over the past twenty years has been to develop anaerobic processes that retain biomass in a variety of forms such that the SRT can be increased while the HRT is decreased.

The goal has been to retain, rather than waste the biocatalyst (bacterial consortia) responsible for the anaerobic process. As a result of this effort, gas yields have increased and digester volumes decreased. A measure of the success of biomass retention is the SRT/HRT ratio. In conventional digesters, the ratio is 1. 0. Effective retention systems will have SRT/HRT ratios exceeding 3. 0. At an SRT/HRT ratio of 3. 0 the digester will be 1/3rd the size of a conventional digester.

Digester Loading (kg/m^3/d)

Neither the hydraulic retention time (HRT), nor the solids retention time (SRT) tells the full story of the impact that the influent waste concentration has on the anaerobic digester. One waste may be dilute and the other concentrated. The concentrated waste will produce more gas per gallon and affect the digester to a much greater extent than the diluted waste. A more appropriate measure of the waste on the digester's size and performance is the loading. The loading can be reported in pounds of waste (influent concentration x influent flow) per cubic foot of digester volume. The more common units are kilograms of influent waste per cubic meter of digester volume per day ($kg/m^3/d$). One ($kg/m^3/d$) is equal to 0.0624 ($lb/ft^3/d$). The digester loading can be calculated if the HRT and influent waste concentration are known.

The loading in ($kg/m^3/d$) is simply:

Where C_I is the influent waste concentration in grams. Increasing the loading will reduce the digester size but will also reduce the percentage of volatile solids converted to gas.

Food to Microorganism Ratio

The food to microorganism ratio is *the key factor* controlling anaerobic digestion. At a given temperature, the bacterial consortia can only consume a limited amount of food each day. In order to consume the required number of pounds of waste one must supply the proper number of pounds of bacteria. The ratio of the pounds of waste supplied to the pounds of bacteria available to consume the waste is the food to microorganism ratio (F/M).

This ratio is the controlling factor in all biological treatment processes. A lower the F/M ratio will result in a greater percentage of the waste being converted to gas. Unfortunately, the bacterial mass is difficult to measure since it is difficult to differentiate the bacterial mass from the influent waste. The task would be easier if *all* of the influent waste were converted to biomass or gas.

In that case, the F/M ratio would simply be the digester loading divided by the concentration of volatile solids (biomass) in the digester (L/C_d). For any given loading, the efficiency can be improved by lowering the F/M ratio by *increasing the concentration of biomass in the digester*. Also for any given biomass concentration within the digester, the efficiency can be improved by decreasing the loading.

Unfortunately, a portion of the influent waste is not processed or converted to biomass or gas by the bacteria. In that case the F/M ratio is equal to the VS loading divided by the digester VS measured (VSD) minus the unprocessed Volatile Solids (VSUP). The unprocessed volatile solids may include refractory or non-degradable biological products produced by the bacteria.

End Product Removal

The end products of anaerobic digestion can adversely affect the digestion process. Such products of anaerobic digestion include organic acids, ammonia nitrogen, and hydrogen sulfide. For any given volatile solids conversion to gas, the higher the influent waste concentration, the greater the end product concentration. End product inhibition can be reduced by lowering the influent waste concentration or by separately removing the soluble end products from the digester through elutriation. Elutriation is the process of washing the solids (bacteria) with clean water to remove the products of digestion. The contact process provides an efficient means of removing the end products of digestion. End product removal can be enhanced by elutriation, which is easily incorporated into the contact process.

Digester Types

A vast array of anaerobic digesters have been developed and placed in operation over the past fifty years. A variety of schemes could be used to classify the digestion processes. For dairy waste, the most important classification is whether or not it can be used to convert dairy waste solids to gas while meeting the goals of anaerobic digestion.

The goals of dairy waste anaerobic digestion are as follows:

- Reduce the mass of solids
- Reduce the odours associated with the waste products
- Produce clean effluent for recycle and irrigation
- Concentrate the nutrients in a solid product for storage or export
- Generate energy
- Reduce pathogens associated with the waste

In addition, the digester must be able to handle or process the dairy waste stream. Dairy waste is a semi-solid slurry. Much of the energy value is in the solids. Consequently, the process must be able to convert solids to gas without clogging the anaerobic reactor. The process must also be able to handle bedding material, sand and other foreign materials associated with typical dairy waste. In addition, if the dairy manure is a dilute waste, the process must be capable of mitigating stratification and solids separation within the reactor.

Processes that are not Appropriate for Digesting Dairy Manure

A variety of high rate anaerobic processes, which retain bacteria have been developed to treat soluble organic industrial wastes. These "high rate" digesters have reduced hydraulic detention times from 20 days to a few hours. They include anaerobic filters, both upflow and downflow, and a variety of biofilm processes such as fixed film packed bed reactors. Bacteria are retained in these reactors as films on carriers such as plastic beads, or sand, or on support media of all configurations. The waste washes past the retained bacteria.

The bacteria convert the soluble constituents to gas but have little opportunity to hydrolyze and degrade the particulate solids, unless the solids become attached to the biomass. These reactors are *not* suitable for digesting dairy waste since they are not effective in converting particulate solids to gas and tend to clog while digesting dairy manure slurries. These high rate reactors can treat the soluble component of dairy waste. But only a fraction of the available energy will be recovered.

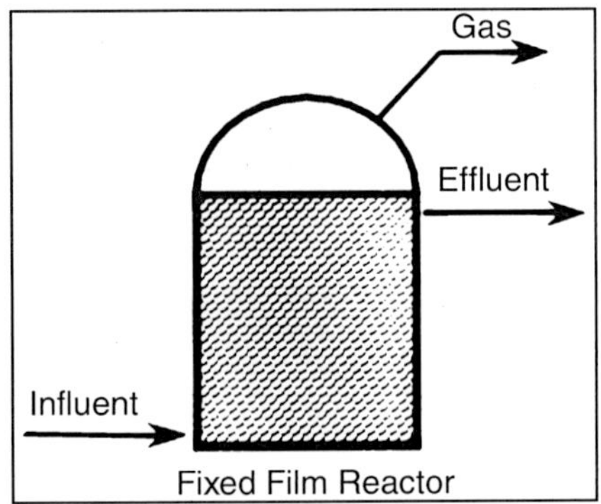

Fig. Packed Fixed Film Reactor

A widely used industrial waste anaerobic digester is the UASB or "Upflow Anaerobic Sludge Blanket", reactor. The process stores the anaerobic consortia as pellets, approximately the size of a pea. The upflow anaerobic sludge blanket reactor (UASB) is widely used in industrial treatment processes throughout the world.

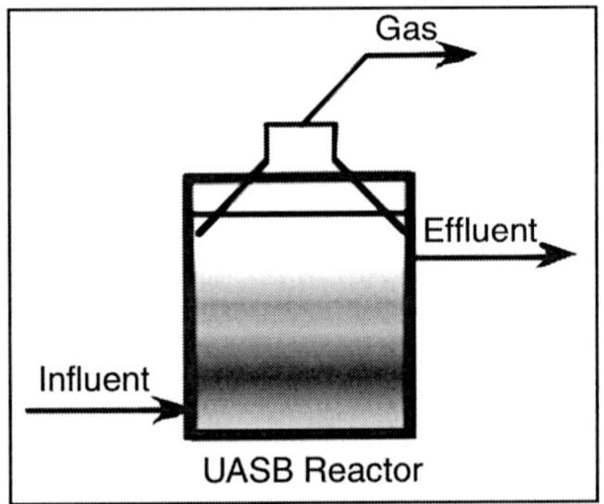

Fig. Upflow Anaerobic Sludge Blanket Reactor

It is an extremely effective process for converting soluble organic materials, such as sugar to methane gas. It has not been used for processing dairy waste since it is ineffective in converting solids to gas. It is primarily used to convert non-particulate or soluble waste to gas.

The anaerobic baffled reactor is a horizontal version of the upflow anaerobic sludge blanket reactor. Both store large quantities of anaerobic bacteria as pellets approximately the size of a pea. Unfortunately, these very successful anaerobic reactors are not effective in digesting particulate waste.

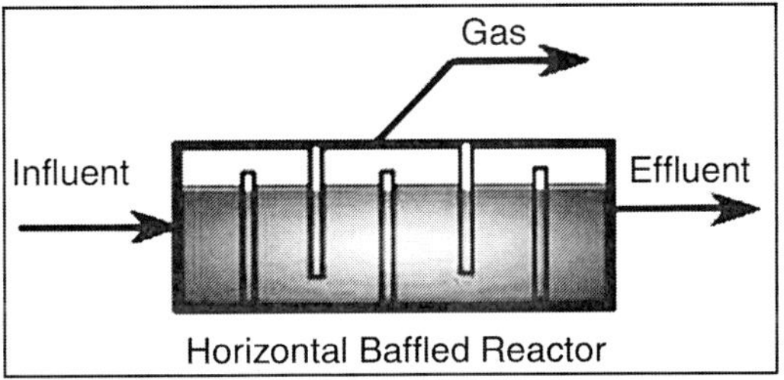

Fig. Baffled Reactor

Particulate solids tend to settle in the horizontal baffled reactor (HBR) while organic fibres will form a mat on the surface. There are no known instances of the HBR being used for the treatment of dairy waste. Unless the dairy waste was thoroughly screened and all particulate matter removed the HBR would tend to become clogged. The removal of solids by screening and gravity sedimentation will eliminate up to 80 per cent of the energy generating potential from dairy waste.

Processes that can be used for Digesting Dairy Manure

The processes that have been used for digesting dairy waste can be subdivided into high rate and low rate processes. Low rate processes consist of covered anaerobic lagoons, plug flow digesters, and mesophilic completely mixed digesters. High rate reactors include the thermophilic completely mixed digesters, anaerobic contact digesters, and hybrid contact/fixed film reactors.

Anaerobic Lagoons (Very Low Rate)

Anaerobic lagoons are covered ponds. Manure enters at one end and the effluent is removed at the other. The lagoons operate at psychrophilic, or ground temperatures. Consequently, the reaction rate is affected by seasonal variations in temperature.

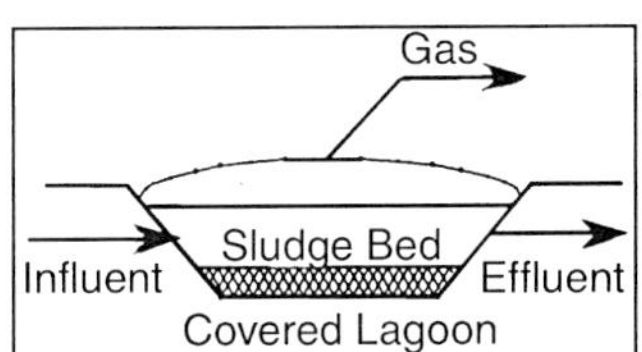

Fig. Anaerobic Lagoons

Since the reaction temperature is quite low, the rate of conversion of solids to gas is also low. In addition, solids tend to settle to the bottom where decomposition occurs in a sludge bed. Little contact of bacteria with the bulk liquid occurs. The biomass concentration is low, resulting in very low solids conversion to gas (High F/M ratio with poor growth rates at low temperatures). Little or no mixing occurs. Consequently, lagoon utilization is poor.

Anaerobic lagoons have been used to treat parlor and free stall flush water. Gas production rates have been low and seasonal. Solids may be screened and removed prior to entering the lagoon. A considerable amount of energy potential is lost with the removal of particulate solids. The advantage of anaerobic lagoons is the lowcost. The low cost is offset by the lower energy production and poor effluent quality. Periodically the covered lagoons must be cleaned at considerable cost. Nuisance odours may be generated while cleaning the lagoons.

Completely Mixed Digesters (Low Rate)

The most common form of an anaerobic digester is the completely mixed reactor. Most sewage treatment plants and many industrial treatment plants use a completely mixed reactor to convert waste to gas. The completely mixed reactor is a tank that is heated and mixed. Most completely mixed reactors operate in the mesophilic range. All of the initial anaerobic digesters used to treat dairy manure were completely mixed mesophilic digesters.

The cost of mixing is high, especially if sand, silt, and floating materials, present in the waste stream, must be suspended throughout the digestion period. Some completely mixed reactors operate in a thermophilic range where sufficient energy is available to heat to reactor. Highly concentrated readily degradable waste is required in order to generate sufficient heat for the thermophilic range of operation. Completely mixed thermophilic digesters are used in the EEC to treat animal manure. Recently, completely mixed thermophilic digesters were proposed in Oregon to treat dairy manure.

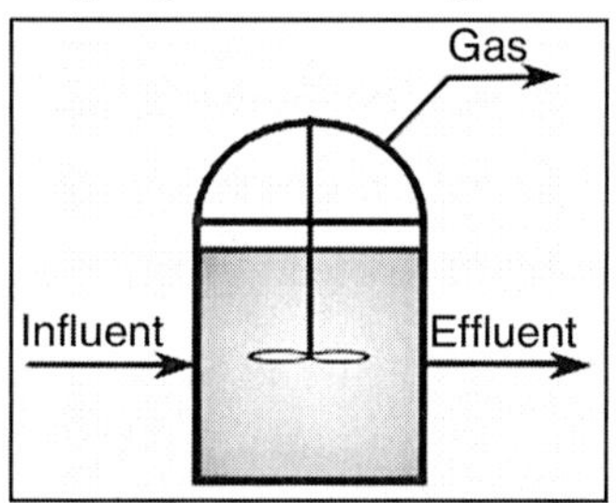

Fig. Completely Mixed Reactor

Most completely mixed reactors are heated with spiral flow heat exchangers. These heat exchangers apply hot water to one side of the spiral and the anaerobic slurry to the other. The spiral heat exchangers have proven to be a successful method of efficiently transferring heat. Completely mixed reactors can be constructed of a variety of materials. In the U.S. most completely mixed reactors have a low profile with a diameter greater than the height.

Some municipal digesters in the U.S., and most in Europe have an egg shape with a height much greater than the diameter. The egg shape enhances mixing while eliminating much of the stratification. Completely mixed reactors can have fixed covers, floating covers, or gas holding covers. Most municipal digesters have floating covers. Floating covers are more expensive than fixed covers. Mixing can be accomplished with a variety of gas mixers, mechanical mixers, and draft tubes with mechanical mixers or simply recirculation pumps.

The most efficient mixing device in terms of power consumed per gallon mixed is the mechanical mixer. Most municipal digesters are intensely mixed to reduce the natural stratification that occurs in a low profile tank. A large amount of evidence has been accumulated over the past 10 years indicating that intense mixing may inhibit the bacterial consortia. But, intense mixing is required to keep sands and silts in suspension. The advantage of the completely mixed reactor is that it is a proven technology that achieves reasonable conversion of solids to gas. It can be applied to the treatment of slurry waste such as dairy manure.

The disadvantage of the completely mixed reactor is the high cost of installation, and the energy cost associated with mixing the digester. The completely mixed conventional anaerobic digester is a biomass growth based system. The process requires a constant conversion of a portion of the feed solids to anaerobic bacteria rather than gas. Since anaerobic bacteria are constantly wasted from the process, new bacteria must be produced to replace the lost bacteria. If the bacteria are retained, the portion of the waste that would have been converted to new bacterial cells will be converted to gas. For this reason, bacterial growth based systems are not as efficient as retained biomass systems. The advantage of the completely mixed thermophilic reactor is the rapid conversion of solids to gas and biomass. Some claim that the rate of conversion is three times greater with thermophilic reactors.

Consequently, the HRT can be lower and the gas production greater. The disadvantage of the thermophilic reactor is the energy required to heat cold dairy manure to thermophilic temperatures. Additional costs are incurred in tank insulation and heat exchangers. Sufficient heat may not be available from the gas produced unless the solids are highly concentrated. Thermophilic digestion of dairy manure cannot be used with manure diluted with parlor or flush water since sufficient energy will not be available to meet the heat requirements. In addition, the higher temperature thermophilic reactors increase end product inhibition, especially ammonia and organic acids.

Plug Flow Digesters (Low Rate)

The plug flow anaerobic digester is the simplest form of anaerobic digestion. Consequently, it is the least expensive. The plug flow digester can be a horizontal or vertical reactor. The waste enters on one side of the reactor and exits on the other. Since bacteria are not conserved, a portion of the waste must be converted to new bacteria, which are subsequently wasted with the effluent. Since the plug flow digester is a growth based system, it is less efficient than a retained biomass system. It converts less waste to gas.

Plug flow systems are subject to stratification wherein the sands and silts settle to the bottom and the organic fibres migrate to surface. The stratification can be partially inhibited by maintaining a relatively high solids concentration in the digester. Periodically, solids must be removed from the plug flow reactor. Since there is no easy way of removing the solids, the reactor must be shut down during the cleaning period. The cost of cleaning can be considerable. Since the solids concentration must be maintained at high levels, dilute milk barn waste is normally excluded from the digester. Plug flow reactors are normally heated by a hot water piping system within the reactor. The hot water piping system can complicate the periodic cleaning of the reactor.

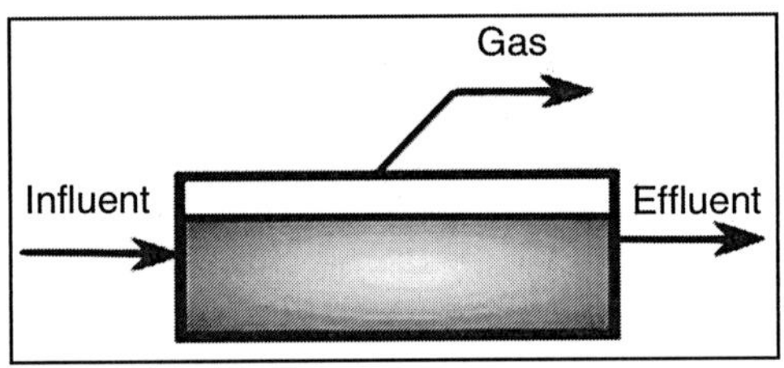

Fig. Plug Flow Reactor

The plug flow reactor is a simple, economical system. Applications are limited to concentrated dairy manure containing a minor amount of sand and silt. If stratification occurs because of a dilute waste or excess sand, significant operating costs will be incurred.

Contact Digesters (High Rate)

The contact reactor is a high rate process that retains bacterial biomass by separating and concentrating the solids in a separate reactor and returning the solids to the influent. More of the degradable waste can be converted to gas since a substantial portion of the bacterial mass is conserved. The contact digester can be either completely mixed or plug flow. It can be operated in the thermophilic or mesophilic range. The contact reactor can treat both dilute and concentrated waste provided the separator can concentrate the digester effluent solids sufficiently to enhance the process.

A wide variety of separators have been tested over the past 30 years. Initially gravity separators (settling tanks), or solids thickeners were used. It was soon discovered that the solids could not be sufficiently concentrated in a gravity separator without degassing to remove the gas bubbles attached to the solids. Actively fermenting digester effluent containing gas bubbles floated rather than settled in the separator.

Lamella or plate separators have also been used to concentrate the biomass after degassing. Both of these gravity-settling techniques are not effective for concentrated digester solids. Gravity separation techniques are effective with dilute waste following a completely mixed reactor. Separation requires several days of detention. The completely mixed digester will prevent stratification. The effluent is then allowed to separate by gravity in the separation reactor.

The digester solids concentration should be less than 2.5 per cent for gravity separation to be used. Long separator detention times are required. Mechanical separation devices have been tested to reduce the detention time required by gravity separation. Centrifuges, gravity belts, membranes, and other mechanical separators have been used with limited success. These disruptive devices have been shown to inhibit the bacterial consortia and thus limit the effectiveness of the contact process. Burke used gas flotation to separate and concentrate the digester effluent for the efficient and tranquil recovery of the anaerobic consortia.

The process has been used for the digestion of dairy manure, sewage sludge, and potato waste. It has been shown to be effective for concentrating the biomass from actively fermenting digester liquors without the need for degassing.

The process has been referred to as the AGF or "anoxic gas flotation" process. Gas flotation can achieve significantly greater biomass concentrations than gravity separation without the adverse consequences associated with disruptive separation techniques.

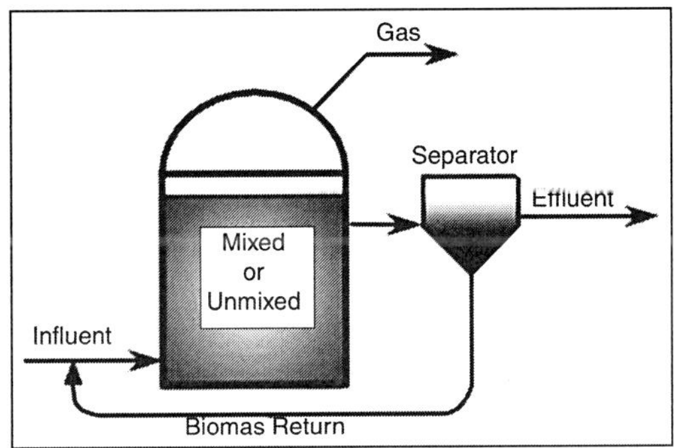

Fig. Contact Reactor

Gas flotation can also remove enzymes, organic acids, and other products of digestion that cannot be removed through settling or other mechanical means. Finally, gas flotation can be performed in a non-mechanical manner, which is simple to operate and maintain. During the contact process, refractory organic and inorganic solids accumulate within the system. The accumulated sands, silts, and non-degradable organic fibres dictate the rate of solids wasting. Wasting the nonbiodegradable solids causes the loss of bacterial mass and reduced process efficiency. The anaerobic contact process can utilize mechanical separating devices to remove refractory solids from the digestion system.

Sequencing Batch Reactors (High Rate)

A sequencing batch reactor is a contact digester, which utilizes the same tank for digestion as well as separation. In a sequencing batch reactor the same tank is used to digest the waste and separate the biomass from the effluent liquor. Generally, two or more tanks are used. The tanks are operated in a fill and draw mode. The separation is accomplished by gravity. Consequently, a more dilute, screened waste is treated. Laboratory scale sequencing batch reactors have been used to digest dairy manure.

Contact Stabilization Reactors (High Rate)

The anaerobic contact stabilization process is a more efficient contact process. Burke used the anaerobic contact stabilization process for the digestion of both dairy manure and potato waste.

The process has the advantage of efficiently converting slowly degradable materials such as cellulose in a highly concentrated reactor. Organic materials, which can be degraded rapidly, are digested in the contact reactor. The bacteria and slowly degradable organics are removed and degraded in a highly concentrated reactor.

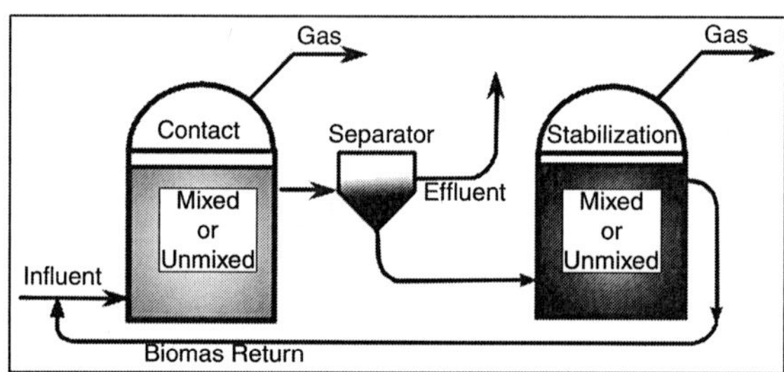

Fig. Contact Stabilization Reactor

Phased Digesters

Both acid phased and temperature phased digestion have been used to convert municipal sludge to gas. Acid phased digestion takes advantage of the fact that the acid forming bacteria have a much higher growth rate than the methanogens. Consequently, the initial reactor can be much smaller than the subsequent methane producing digester. Acid phased digestion offers greater efficiency in the size of the anaerobic digesters.

Acid phased digestion has not been applied to dairy waste. Temperature phased digestion has been applied to the digestion of sewage sludge.

The initial digester is operated in the thermophilic mode followed by a second digester, which is operated in the mesophilic mode. In the first thermophilic digester, pathogens are destroyed. In the second mesophilic digester, the mesophilic bacteria consume the organic acids created in the thermophilic reactor. Consequently, the odours associated with a thermophilic effluent are eliminated while achieving the desired pathogen destruction.

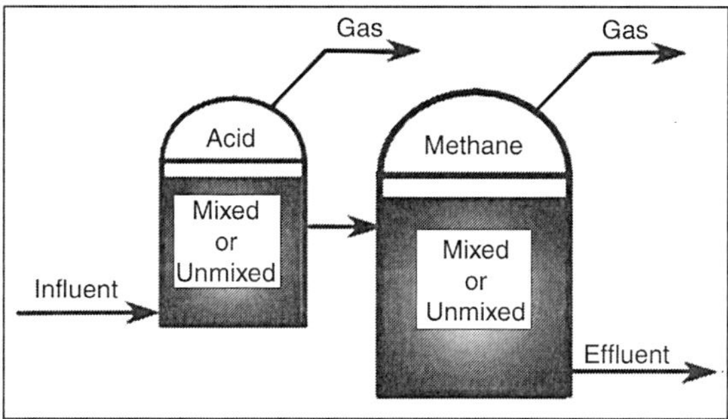

Fig. Acid Phased Digester

Temperature phased digestion has been used to digest dairy manure. In addition it must be pointed out that completely mixed reactors are not completely effective in removing pathogens.

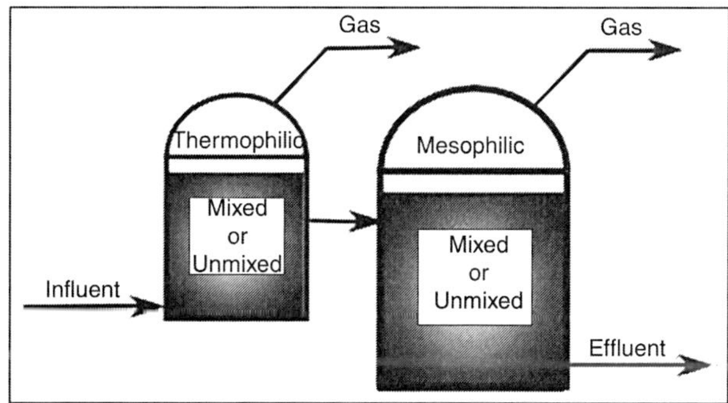

Fig. Temperature Phased Digester

Hybrid Processes

A number of hybrid processes have been developed and applied to many different kinds of waste materials. The hybrid processes incorporate a combination of the previously described configurations.

QUALITATIVE ANALYSIS OF ANAEROBIC PROCESSES

In order to assess the various digester configurations one must define their limitations for dairy waste digestion.

Solids Concentration Limitations

The ability to process a variety of manure concentrations is important. Even though the solids may be collected in a concentrated form, there will be times when the solids become diluted. The inverse is also true. The dilute parlor waste may become concentrated for variety of reasons. The mesophilic and thermophilic completely mixed processes and the contact process can handle a variety of influent manure concentrations. Their operating performance is not limited by the manure concentration. On the other hand the plug flow digester and the anaerobic lagoon are limited by the influent manure concentration. The plug flow digester will stratify at low feed concentrations. The anaerobic lagoon will accumulate non-degraded solids at high influent solids concentrations.

Digestion of the Entire Waste Stream

The thermophilic and mesophilic completely mixed reactors and the plug flow contact process can digest the entire waste stream since neither are limited by the concentration of the influent waste. Plug flow reactors will be able to process the concentrated or scraped manure. Plug flow reactors will not be able to economically process the parlor waste or a mixture of the parlor and scraped waste. The anaerobic lagoon can process primarily liquid waste after the removal of fibres and particulate solids.

The current practice of screening fibres and settling solids to remove particulate matter is not compatible with achieving high energy yields through anaerobic digestion. It is generally accepted that screening will remove at lease 15 per cent of the influent COD. Recent analysis has shown that screening and sedimentation will remove 60 per cent or more of the COD that could be converted to gas. Solids separation should follow, rather than precede anaerobic digestion.

Foreign Material Processing

High concentrations of sand and silt are not compatible with the plug flow digester or the anaerobic lagoon. Completely mixed reactors can operate with minor concentrations of foreign material by maintaining the material in suspension through intense mixing. The contact process incorporating grit removal as described by Burke is not limited by the concentration of foreign material.

Odour Control

Most properly operated anaerobic digesters will eliminate the generation of odours from the site. However, both plug flow anaerobic digesters and the anaerobic lagoon must be periodically cleaned. During the cleaning process odours are generated. The thermophilic completely mixed reactors produce an effluent that is far more odorous than mesophically digested waste. Contact and completely mixed digesters significantly reduce odours and may *not* require cleaning, especially if refractory inorganic and organic solids removal is practiced.

Stability, Flexibility, and Reliability

Each type of anaerobic reactor imposes requirements for its proper operation. The inability to meet those requirements, such as operating temperature, may result in process failure. The mesophilic process is more reliable than the thermophilic process because of the greater risk associated with meeting the thermophilic temperature requirements utilizing a cold waste at cold temperatures.

The anaerobic process has been labeled an unreliable process because of frequent toxic upsets. The contact process is a more reliable process in preventing process failure, foaming, and loss of biomass. Retained biomass systems are the least likely to fail because of a large quantity and diversity of the biocatalyst in the digestion system. The addition of elutriation, or the washing of biological solids to remove inhibitory products, adds further stability to the process.

The contact stabilization process is the least likely to be upset by changes in hydraulic flow, or organic loading. Since mixing is essential to any completely mixed process, mixing failures, or inadequacies may result in poor performance. The plug flow contact process also poses little risk of failure due to mixing inadequacies or solids accumulation in the digester. The complexity of the process will also affect its reliability.

The thermophilic digestion of dairy manure must incorporate a complex heating and heat recovery system. Its reliability will be less than a system that does not have such complexity. The contact process and the contact stabilization process are also more complex systems. They have more of an opportunity to fail. Redundant equipment and robust controls are essential to improving the reliability of complex systems.

Nutrient Concentration and Retention

The process of anaerobic digestion will convert nutrients from an organic form to an inorganic form. In plug flow, completely mixed, and thermophilic reactors the quantity of nutrients entering the reactor equals the quantity of nutrients exiting the reactor. However, in retained biomass digesters such as the contact process, sequencing batch reactors, and fixed film reactors, nutrients may be concentrated in a separate waste solids stream.

Dugba demonstrated that the effluent from a sequencing batch reactor contained less than 50 per cent of the influent phosphorus. The balance of the phosphorus was concentrated in the biosolids. Burke demonstrated the retention and concentration of 90 per cent of the influent phosphorus and 43 per cent of the influent total nitrogen in the waste solids that was only 1/5 of the influent volume. The ability to concentrate nutrients is an important characteristic of the selected anaerobic process since it provides the dairy operator with the control necessary to manage nutrient application to the land.

Additional Substrate Processing

Hobson studied the effect of adding cellulose to dairy manure. His research indicated that the volatile solids conversion to gas would be substantially improved through the addition of cellulose. At a 16-day hydraulic retention time the volatile solids conversion to gas increased from 30 per cent with no cellulose to 51 per cent with a manure containing six per cent cellulose. Many commercial digesters supplement the influent with food waste or food processing waste.

Collection of tipping fees improves the economic viability of anaerobic facilities. The Tillamook project in Oregon proposed to supplement the influent waste with municipal solid waste to increase revenues. The proposed Myrtle Point project in Oregon may treat milk-processing waste to increase revenues. The ability to treat a wide variety of influent substrates, and thereby enhance the economics, is an important process characteristic. The completely mixed and contact processes can process a variety of added substrates.

Energy Production

The quantity of energy produced from each gallon of waste processed is strictly a function of the percentage conversion of volatile solids to gas. Each pound of volatile solids destroyed will produce 5.62 cubic feet of methane. Each cubic foot of methane will contain 1000 Btu's of energy. Therefore, each pound of volatile solids converted will produce 5620 Btu's of energy.

At 35 per cent conversion efficiency, each pound of volatile solids destroyed will produce 0.58 kWh of energy. It is therefore important to look at the conversion efficiencies of the various anaerobic processes. As pointed out earlier, the conversion of volatile solids to gas is a function of the organic loading to the digester. Higher percentage conversions to gas are achieved at lower organic loadings. Low loadings however, translate into larger digestion facilities.

However, it is possible to achieve a higher volatile solids conversion to gas by increasing the digester loading while maintaining a higher biomass concentration in the digester. In other words, the food to microorganism (F/M) ratio remains low resulting in a higher rate of conversion. The rate of volatile solids conversion to gas is related to the type of anaerobic digester used.

Conventional completely mixed and plug flow digesters, which do not retain biomass, should have comparable volatile solids destructions. Anaerobic lagoons will have a lower rate of conversion, while high rate retained biomass reactors will have higher rates of solids conversion to gas. Each is discussed separately below.

Conventional Digesters

A review of recent dairy waste anaerobic digestion studies has established that most engineers anticipate a 50 per cent conversion of volatile solids to gas. The planned Three-Mile Farm (Oregon) dairy waste *thermophilic* anaerobic digestion facility is expected to achieve a 50 per cent volatile solids conversion to gas. The C. Bar M. (Idaho) *plug flow* anaerobic digester facility anticipated a 50 per cent conversion of dairy waste volatile solids to gas.

The recently completed Myrtle Point (Oregon) feasibility study utilizing the *gravity separation contact process* anticipated a 50 per cent conversion of dairy waste volatile solids to gas. Relatively high loading rates were anticipated in each case. The organic loading rates varied between 5.6 and 6.4 kg per cubic meter per day. The available literature does not support such high volatile solids conversions to gas at high organic loading rates.

A summary is as follows:

- "The Monroe Honour Farm *completely mixed anaerobic digester* achieved a maximum of 40 per cent volatile solids conversion to gas at a loading rate of 6 kg/m^3/d. Jewel operated a *plug flow anaerobic digesters* at an organic loading rate of 2.37 and achieved a 32.4 per cent conversion to volatile solids to gas."

Converse operated both thermophilic and mesophilic completely mixed anaerobic digesters at a loading rate of 4.2 kg/m^3/d. Both thermophilic and mesophilic digesters achieved a 41 per cent conversion of volatile solids to gas. Bryant on the other hand, operated *completely mixed thermophilic* digesters at loadings of 6.5 to 10.78 kg/m^3/d and achieved 50 per cent volatile solids conversion to gas. Recently, Ahring reported a 28 per cent volatile solids conversion in a thermophilic digester operated at a loading of 3 kg/m^3/d.

Ghaly operated a dairy waste completely mixed mesophilic digester at a loading of 3.6 kg/m^3/d. He achieved a 46 per cent conversion of volatile solids to gas. Qasim operated a completely mixed mesophilic digester at an organic loading rate of 3.2 kg/m^3/d and achieved a 52.9 per cent volatile solids conversion to gas. Echiegu operated a completely mixed dairy waste digester at an organic loading rate of 2 kg/m^3/d but only achieved a 40 per cent conversion.

Robbins also operated a completely mixed mesophilic digester at an organic loading rate of 2.6 kg per cubic meter per day that achieved a 30 per cent conversion of volatile solids to gas. Hills and Kayhanian operated a completely mixed mesophilic digester at a 1.8 kg/m^3/d loading that achieved a 31 per cent volatile solids destruction and a 38 per cent conversion at 1.0 kg/m^3/d.

On the other hand, Pigg operated a completely mixed mesophilic anaerobic digester at an organic loading rate of 1.0 kg/m^3/d and achieved a peak volatile solids conversion to gas of 64 per cent. As can be observed the published literature values are highly variable. The results generally confirm Smith's conclusion that mesophilic digesters can achieve a 40 per cent conversion of volatile solids at a loading of 5.7 kg/m^3/d.

Better conversions can be achieved at lower loadings. Thermophilic reactors appear to achieve greater conversions at high loadings while mesophilic reactors appear to achieve greater conversions at lower loadings. Lusk provided information on the performance of full-scale plug flow and completely mixed anaerobic digesters treating dairy manure. The loading and per cent volatile solids conversion can be calculated from the information he presented. Figure below presents the results of the analysis of the Lusk data.

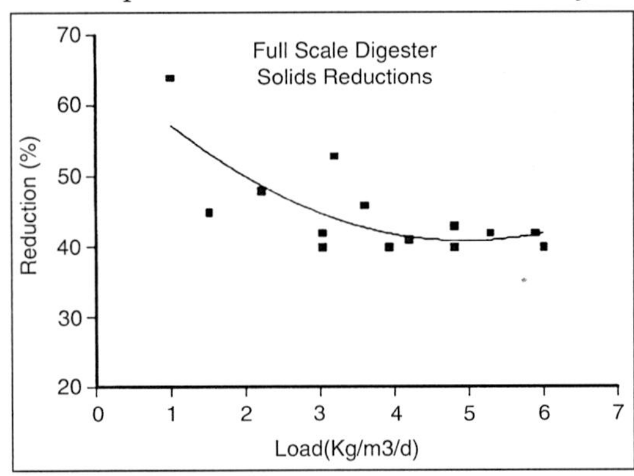

Fig. Full Scale Mesophilic Digester vs. Reductions

Lagoons

California Polytechnic State University in San Luis Obispo constructed an anaerobic lagoon to treat flush waste from a 350 animal dairy. The screening system removed 15 per cent of the manure volatile solids. The lagoon was projected to achieve a 35 per cent volatile solids conversion to gas at a loading of 0.04 kg/m^3/d.

High Rate Anaerobic Reactors

Wilkie reported the use of a fixed film reactor treating screened dairy manure having the volatile solids concentration of 3.2 g per liter. The fixed film reactor achieved a 57 per cent COD conversion to gas. The gas contained 78% methane. On a COD basis, the loading was approximately 10.7 kg/m^3/d.

The hydraulic retention time was three days. Unfortunately, most of the potential gas production was lost in the screened and settled manure that was not processed by the fixed film digester. Dugba reported on the use of a temperature phased anaerobic sequencing batch reactor system treating dairy manure.

The systems were operated at a three-day hydraulic retention time. The dairy manure was screened utilizing a 2 by 2 mm screen opening. The biogas had a methane concentration of 62 to 66%. The sequencing batch reactors reportedly had a SRT/HRT ratio of 3 to 4. (Observation of phosphorus retention indicated that the systems were not very effective in retaining biomass.)

The systems were operated at a three-day HRT. The volatile solids feed concentration varied between 0.6 and 1.2 per cent volatile solids. The loadings ranged from 2 to 4 $kg/m^3/d$. The volatile solids conversions to gas ranged from 30 to 41 per cent. Umetsu reported on the performance of a horizontally baffled anaerobic digester treating dairy manure at ambient temperatures. The average HRT was ten days. The methane content of the biogas was 58 per cent. The digester achieved a 25 per cent volatile solids reduction at a volatile solids loading of 7.3 $kg/m^3/d$. Wuhou reported on the anaerobic digestion of dairy manure from a 2900 cow dairy. The digester was a mesophilic up flow mixing reactor operated at a volatile solids loading of 3.7 kg per cubic meter per day. The digester achieved a 78.2 per cent volatile solids conversion to gas. Burke operated a mesophilic AGF contact reactor at an average loading of 2.13 kg per cubic meter per day. The average COD conversion to gas was 66. 3 per cent. The maximum COD conversion to gas was 69% at a loading of 2.4 kg per cubic meter per day. The minimum conversion was 59.2 per cent during the startup period. The biogas had a methane concentration between 60 and 65 per cent. A summary of the expected performance for each type of anaerobic reactor is presented in the table below.

Table. Expected Percentage vs Conversion to Gas

Process	Load	Conversion to Gas
Entire Waste Stream		
Completely mixed mesophilic	High	35 to 45%
Completely mixed thermophilic	High	45 to 55%
Contact	High	50 to 65%
Partial Waste Stream		
Plug mesophilic	High	35 to 45%
Fixed film	High	55 to 65%
Lagoon	Low	35 to 45%

Cost of Anaerobic Processes for Dairy Waste

The cost of a dairy waste management system can be subdivided into the following elements:

- *Housing:* Determines the percentage of manure actually collected
- *Collection*: A means of collecting the waste by manual, or automatic scraper, vacuum truck, or flush.
- *Pre-processing*: Screening and or sedimentation prior to digestion
- *Anaerobic Digestion:* The solids conversion process
- *Post-processing:* The concentration of solids after digestion
- *Energy Production:* Engine Generator or Turbine with heat recovery
- *Liquid Handling and Irrigation:* The storage and disposal of liquid waste
- Solids Disposal

As pointed out earlier, use of any particular system will have an effect on the other. For example, if a flush system is used the anaerobic digester must be larger. If pre screening and sedimentation are used, the amount of energy produced will be lower. Post processing will establish the cost of ultimate solids disposal. In many cases, solids must be exported from the site. The use of an anaerobic digestion system may eliminate the need for pre-processing or screening and sedimentation.

The cost of a complete dairy waste management system may exceed $1,200 per cow. Anaerobic digestion system costs however, are confined to the cost of the anaerobic process, post solids handling, and energy production. A review of the anaerobic digestion system costs at U.S. dairies compiled by Lusk has established that the typical anaerobic system constructed in the U.S. had an average cost of $470 per cow.

The proposed thermophilic digestion project at Three Mile Farm (21,000 cows) in Oregon projected a cost of $710 per cow. The proposed contact process at Myrtle Point (4,500 cows) Oregon has a proposed cost of $678 per cow for digestion, solids handling, and power generation. The recently constructed Cal Polly flush system anaerobic lagoon had a cost of $800 per cow.

Table. Adjusted Capital Costs

System Cost	% Treated $/Cow	Reported Cost $/Cow	Adjusted
Lusk U.S. Average	85	$470	$552
Three Mile Farm Thermo	80	$710	$887
Myrtle Point Oregon	80	$678	$847
Cal Poly (lagoon)	100	$800	$800

Anaerobic systems for digestion, solids processing, and generation are expected to cost $500 to $800 per cow. The per-cow capital cost estimates can be deceptive since some processes treat the entire waste stream while others treat only a portion of the waste stream. For example, the plug flow systems documented by Lusk treat only the concentrated portion of the manure while excluding the milk parlor waste (15% of dairy manure). The table above presents the corrected capital costs for the entire waste stream (100% of cow manure).

Even the adjusted capital cost per cow does not tell the full economic story. Some systems are far more efficient than others in producing power and sequestering nutrients. The best approach is to report the capital costs in terms of dollars per MBtu generated during the first year of operation or dollars per net kW of power sale capacity.

There is little data available in those units for US systems. The costs of a wide variety of European systems have been reported in those units. The capital costs of European systems vary from country to country. Germany produces anaerobic digesters for the least cost per gallon while the Danish systems produce greater amounts of biogas per pound of solids introduced to the digester. German digester systems are constructed for an average cost of $1.52 per gallon. Danish digesters have a capital cost of $50 per GJ or $5.26 per annual biogas therm (100,000 Btu or 1.06 GJ). Power is produced for a capital cost of $10,000 per kW of export capacity. The table below presents the capital and operating cost of European systems for a large facility and a small on farm system.

Table. Capital and Operating Costs of European Digestion Systems

	Large 1 MW 5000 Cow Facility	Small 25 kW 125 Cow Farm
Capital Cost	$9,113,000	$500,000
Annual Operating Cost	$643,000	$8,800
Power Sale Rate $/kW	$0.06	$0.06
Heat Sale $/kW	$0.01	$0.01
Solids Sales	$700,000	$20,000

It must be noted that the capital and operation and maintenance costs are considerably greater in Europe than those reported the US. On the other hand, income derived from the sale of the solids is considerably greater in Europe. The capital cost of dairy waste systems in Idaho are expected to be from $2,700/kW to $6,000/kW exclusive of sales tax, power connection to the site and financing costs (27 to 60% of the European cost).

ALTERNATIVE WASTE MANAGEMENT SYSTEMS

A Dairyman has many choices in designing and operating in modern dairy. Figure presents the basic options for housing cows, the collection of manure, the pretreatment and treatment of manure, the post treatment and final disposal. Each of the options must be judged by the goals of the Dairymen to maintain animal health, recover nutrients, minimize odours, produce energy, and reduce operation and maintenance costs. The following paragraphs summarize the advantages and disadvantages of each alternative. The better alternatives are shaded in the accompanying figure.

EXISTING MANURE HANDLING

Existing manure handling consists of screening, and gravity separating the solids for subsequent composting. The compost produced by existing systems can either be exported or used for bedding material. Up to 70 per cent of the phosphorus nutrients can be diverted from the farm through the use of the existing solids removal processes. However, the conventional solids handling process does not reduce the quantity of material to be handled. It simply separates the material into a solid fraction that can be stacked and hauled and a liquid fraction that can be placed in a holding pond for the required 180-day detention time with a minimum of sedimentation.

The system does not control odours since the liquid fraction containing large quantities of organic acids is discharged to an open lagoon for further decomposition. The system does not produce any energy. Large quantities of energy are consumed in separating the solids and subsequent composting. The operation and maintenance costs are high but the capital costs are low. Overall the commonly used manure handling system does not pay for self nor meet the environmental goals of the Dairymen.

HOUSING

This type of housing establishes the quantity of manure that can be collected economically. Free stall barns permit the collection of 85 per cent of the manure. The remaining 15 per cent is normally deposited in the milk parlor where it is collected through a flush system. In corral systems only 40 to 50 per cent of the manure can be collected from the feed lanes, which are either scraped or flushed.

Manure deposited in the open lot portion of the corral is normally collected once or twice a year. While in the open lot, the manure is degraded both aerobically and anaerobically depending on the moisture content. The degradation process produces odours and greenhouse gases that are discharged to the atmosphere. There is little net energy available from the manure after it has remained in the open lot for 8 to 12 months.

The free stall system is better for animal health since it provides the greatest separation between manure and cow. Since the free stall system provides an opportunity to collect the maximum amount of manure it also provides the opportunity to recover most of the phosphorus nutrients and generate the most energy. *Free stall barns are more expensive to construct than corral or open lot systems. The operation and maintenance cost of the free stall barn is less than corral or open lot systems. Overall the free stall system provides the best manure processing option.*

COLLECTION

Manure can be collected by flush, scrape, or vacuum collection. Scrape and vacuum collection systems have a higher capital and operation and maintenance costs. Flush systems have the lowest capital and operation and maintenance costs. Flush systems also remove substantially all of the manure. The vacuum and scrape systems do not clean the barns as efficiently as a flush system. Flush systems significantly increase the quantity of waste that must be processed through an energy recovery system. The increased quantity of cold, dilute manure can result in much larger treatment facilities and lower temperatures within the anaerobic digesters. Flush systems have also been associated with severe odour problems. Wet flush aisles promote bacterial activity leading to organic degradation and the generation of odours. The large quantity of untreated wastewater that is discharged to open ponds is an additional source of odorous degradation products.

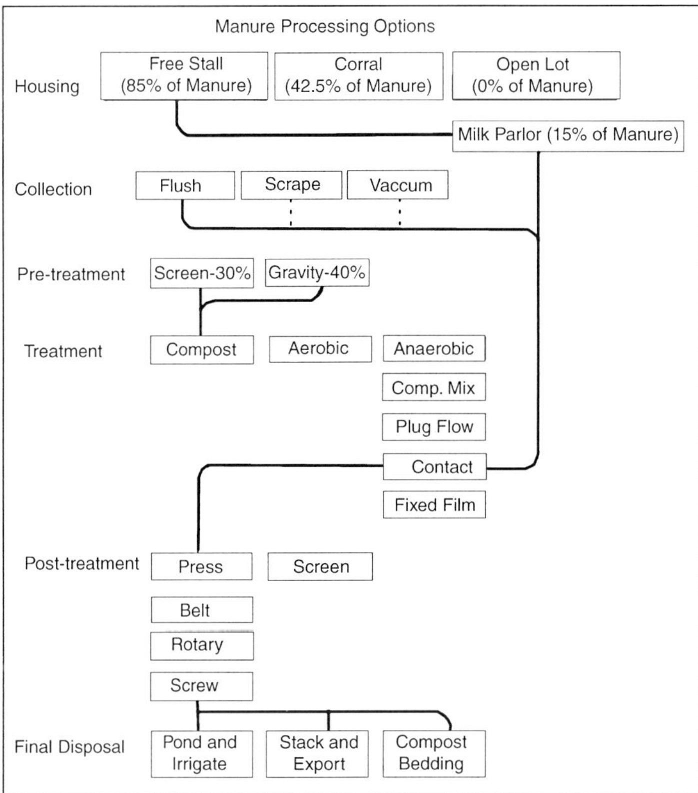

Fig. Manure Processing Alternatives

The problems associated with flush systems can be mitigated to a great extent by flushing once a day. By flushing the aisles once a day the volume of flush water will be significantly reduced such that it can be heated and effectively treated with the anaerobic contact process. Flushing once a day will also provide the opportunity for the aisles to dry, eliminating the generation of odours from open-air microbial degradation.

Scrape and vacuum systems provide a concentrated waste flow, which minimizes the size of down-stream treatment facilities. The choice of which collection system to use will be based on economics since all collection systems, if properly operated, provide an opportunity to recover nutrients, enhance animal health, and minimize odours.

Scrape and vacuum systems have a higher capital, and operation and maintenance costs. The downstream treatment costs however are lower. Flush systems have a lower capital and operating cost but the downstream treatment costs are higher.

TREATMENT

A wide variety of pretreatment and treatment options exist. Typically pretreatment consists of screening and gravity separation of the solids. The recovered solids are allowed to drain and may be subsequently composted for animal bedding. Stacked solids are applied to land once or twice a year depending on the phosphorus land application limitations. The stacked solids may also be exported from the site. Many nutrient management plans require 100 per cent export of the stacked solids. Unfortunately, pretreatment is not beneficial for energy recovery since a significant portion of the solids that can be converted to gas are removed through the pretreatment activity.

Since existing pretreatment processes produce odour, are expensive to operate and maintain, and severely impact potential energy generation it is recommended that pretreatment not be used before anaerobic digestion. Excess sands, silts, and fibres should be removed as part of the anaerobic digestion process. Aerobic treatment is an effective alternative for reducing odour. Aerobic treatment consumes large quantities of energy and has higher operating and maintenance costs. Aerobic treatment however, has lower capital costs than anaerobic digestion, and it is less effective in recovering nutrients than anaerobic digestion.

Anaerobic digestion is the most beneficial treatment option. The contact process is the most effective anaerobic treatment process. Both the fix film and plug flow processes are concentration limited. The contact process can handle a wide variety of solids concentrations. All of the manure from the milk barn to the free stall can be processed through the contact process. In addition to processing a larger percentage of manure, a greater percentage of the solids will be converted to energy. It provides greater load flexibility, allowing dairymen to process other waste materials for additional gas production.

The contact process requires little operation and maintenance. It can be an automated process. Both the contact process and completely mixed processes can handle sand and floating fibres. The contact process and completely mixed digester use more energy than the plug flow process. However, a greater percentage of the solids will be converted to energy. The plug flow process is less expensive than either the contact process or the completely mixed reactor.

The contact process however, uses less energy than the completely mixed reactor and has a much lower capital costs. All mesophilic anaerobic treatment processes are effective in reducing or eliminating odours. All anaerobic processes can sequester most of the nutrients. Since the contact process produces a relatively clear effluent, additional nutrients can be removed. All anaerobic processes produce energy. The contact process will provide the greatest solids retention time leading to a higher energy yield. The plug flow process has the lowest capital costs followed by the contact process. The plug flow reactor also has the lowest operation and maintenance costs. The contact process will have the highest operation and maintenance costs because of the reagents that are used in the biomass separation process. Overall, the contact process offers a greatest benefit.

POST-TREATMENT

The final product from the treatment process consisting of undigested solids, biomass, and inorganic precipitates must be separated such that the solids containing a majority of the nutrients can be stored, stacked, and exported if required. The post-treatment process fulfills the same need as conventional pretreatment process. However, after passing through an anaerobic digestion process the solids are substantially reduced and the nutrients are concentrated.

A number of options exist for post-treatment. They include screens and a variety of presses. The screw press requires the least amount of operation and maintenance. Consequently, the recommended post-treatment is to pass the digested solids through a screw press, separate the solids from the liquid, and either export the solids or compost them for bedding.

The contact process provides more effective liquid solids separation. If the contact process were used, the liquid from the screw press would be recycled to the contact separator. The clean particle free effluent would be discharged from the contact process separator to the storage pond for irrigation.

FINAL DISPOSAL

The final products consisting of a liquid stream and a solid stream must be disposed in accordance with the nutrient management plan. The liquid stream will contain inorganic nitrogen as ammonia and a small amount of phosphorus. The solid stream will contain organic nitrogen and a vast majority of the phosphorus. Both the solid and liquid streams will be fully stabilized and odorless. The solids can be stacked for export or composted for bedding.

5

Membrane Technology in Dairy Processing

Membrane filtration technologies, such as ultra filtration and reverse osmosis, are capable of the molecular fractionation of fluids. Milk is ideally suited for processing by membrane filtration because it is a fluid consisting largely of water, lactose, butterfat, and protein molecules. Separation at the molecular level means that butterfat, lactose, and protein can be isolated from one other. Through the use of cellulose filters and high pressure pumps, membrane technologies take the two-dimensional concept of the venerable cream separator (*i.e.*, milk in, cream and skim out) into the third dimension and even beyond.

Membrane technologies have brought about substantial change in the dairy industry. However, because of the rapid pace of innovation many new dairy products created by membrane technology have not yet gained effective consumer demand. As David Hettinga, Vice President and Chief Technical Office of Land O'Lakes, stated, "one of the problems with this technology is we have a product or a technology chasing the market". The purpose of this research is to introduce readers to the membrane process and then attempt to assess the consumer demand for a few such new products.

The traditional dairy manufacturing paradigm has been to separate whole milk into cream and skim milk using a centrifuge. The skim milk is then John W. Siebert is an associate professor and Alejandro Lalor is a graduate research assistant, Department of Agricultural Economics, Texas A&M University, College Station, TX 77843-2124. Sung-Yong Kim is a research associate at the Korea Rural Economic Institute. The authors wish to thank the Southwest Dairy Farmers of Sulphur Springs, TX and Texas A&M University for their support of this research. The authors also with to thank two anonymous reviewers as well as Dr. Sefa Koseoglu. often evaporated to produced condensed skim.

Most dairy products are made using various combinations of milk, cream, skim, and condensed skim. The shortfall of this traditional technology approach is that protein and lactose (the main ingredients of skim) are bound to one another. A key value of membrane technology is that it enables a separation of these two ingredients. With membrane technology, protein, butterfat, and lactose can be used to manufacture dairy products more directly. Should sales of milk increase due to the development of new products, total dairy-farmer income would be likely to increase as well. With approximately 590,000,000 pounds of nonfat dry milk currently in government warehouses, research into demand expansion remains a high priority for dairy farmers.

TECHNOLOGY REVIEW

The most widely accepted dairy applications of membrane technology have been cost-reducing in nature. For example, most modern cheese plants use membrane technology to extract valuable protein isolates from the whey stream. Whey-protein concentrate is currently an important source of income to all large cheese makers. The portion of a modern cheese plant devoted to whey-product manufacturing and storage can be almost as large as that devoted to cheese. Due solely to the ability of membrane technology to extract protein from whey, whey is no longer a disposal problem-it is now a profit center.

In New Zealand, membrane technology is used to produce a powdered dairy product consisting largely of butterfat and protein. Due to its functionality, this ingredient-called dry ultrafiltered milk or milk protein concentrate (MPC)-can be used to make cheese. MPC is imported to the United States for the purpose of boosting cheese-plant yields. In this regard it is a substitute for domestic nonfat dry milk, and for this reason has been viewed as a threat to the U.S. milk price support programme.

Dairy farmers in remote regions of the United States have used membrane technology to reduce raw milk transportation costs. At the farm, ultrafiltration is being used to remove lactose and water from milk. Also at the farm, reverse osmosis is being used to remove water from milk. Membrane technology will likely replace the traditional cheese vat in the future. The traditional cheese vat is a large kettle (*e.g.*, 5,000-gallon capacity) that uses calf rennet, heat, and agitation in order to yield cheese curd and whey from milk.

The curds are then pressed into blocks of cheese and aged. Membrane technology has the potential to produce cheese by molecular separation of lactose from the butterfat and protein. This would allow the design of equipment that accepts milk as an input and produces liquid cheese and whey as outputs. The liquid cheese stream could then be poured into forms for hardening and aging.

The objective of this research was to determine if membrane technology has the potential to create new commercial dairy products for direct purchase by consumers. The key questions investigated concern the capabilities of membrane technology, the economics of producing new consumer products, and the consumer market potential of any such new products. Before new consumer dairy products such as these can be found and evaluated, the technology must first be understood. Figure provides a membrane technology diagram.

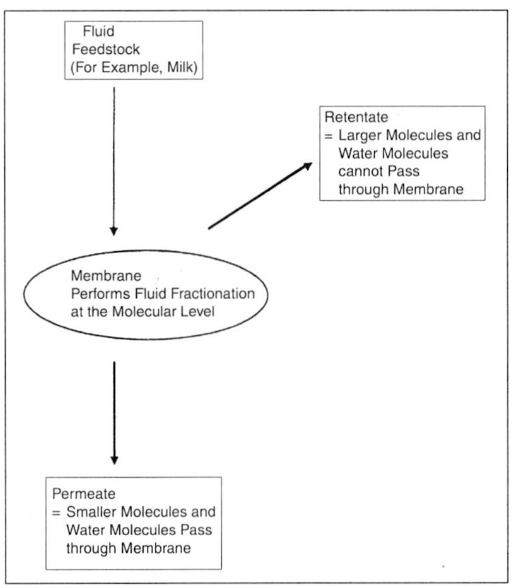

Fig. The Membrane Filtration Process is Initiated by a Fluid being Pumped over a Membrane. This Causes Fractionation at the Molecular Level.

This figure shows a fluid being pumped across a membrane under high pressure. Smaller particles pass through the membrane and are termed permeate. Larger particles cannot pass through and are denoted as retentate. The membrane filtration process can be performed at progressive levels of molecular selectivity. Reverse Osmosis (RO) is a term denoting a very fine membrane-filtration process. To understand filtration in its application to dairy, consider that raw milk consists largely of water, lactose, butterfat, protein, and minerals.

Applying RO to milk would thus produce a permeate which consists mainly of water and a retentate which consists of water, lactose, butterfat, protein, and minerals. Ultrafiltration (UF) allows somewhat larger molecules to pass through the membrane than does RO. In this case, not only water but also lactose molecules will pass through the membrane. Thus, applying UF to milk produces both a permeate consisting of water and lactose and a retentate consisting of water, lactose, butterfat, and protein.

EQUIPMENT INDUSTRY SURVEY

To gain an understanding of the current status of membrane technology in the dairy industry, the authors made an initial survey of the membraneequipment industry. Our objective was to learn about potential new consumer dairy product applications of membrane technology. We contacted nineteen firms involved in various combinations of equipment manufacturing, facilities and/or equipment design, and equipment installation. The authors found these nineteen firms through advertisements in the dairy trade press, through suppliers listed in the International Dairy Foods Association Membership Directory, and through attendees at a Texas A&M University Short Course on Membrane Technology.

The authors do not know what percentage of the dairy membrane manufacturing industry was contacted through their survey but the percentage is believed to be high, as all known firms were contacted. Also, the supplier industry is relatively concentrated. Thus, despite the small number of firms involved, this sample should be considered representative of the dairy membrane equipment industry during 1999. The firms contacted served the entire United States. Nine of the thirteen firms were headquartered in either Minnesota or Wisconsin. Several of the firms were subsidiaries of international companies.

Thirteen of the nineteen firms contacted participated, a response rate of 68 per cent. The responding firms viewed membrane technology as advancing rapidly in terms of fractionation selectivity, methods, and reliability. Technological advances usually originate in Australia, New Zealand, or Western Europe. Consequently, U.S. firms often employ technology after it has proven its value elsewhere. Two dairy industry forces, when taken in combination, likely explain why New Zealand, Europe and Australia have historically taken the lead in the development of membrane technologies.

First, the U.S. Food and Drug Administration has restricted dairy manufacturers from using membrane technology in the production of traditional dairy products such as cheese. The FDA must approve on a firm-by-firm and case-by-case basis that the product manufactured with membrane technology has no compositional or organoleptic differences when compared to a product made in full conformance with regulatory Standards of Identity.

Second, the U.S. pricesupport programme only offers stand-by purchasing authority for cheese, butter, and nonfat dry milk; therefore, membrane-based dairy products such as milk protein concentrate powder would not qualify for the programme. As a result, U.S. dairy-industry investment is often made in traditional production technology in order to reduce exposure to price risk. Manufacturers were asked about the future of membrane technology. The consensus was that byproduct extraction at the dairy processing plant would be the main area for the future impact of membrane technology.

Specifically, eight of twelve manufacturers who responded to a question concerning whether the biggest impact of membrane technology would be at the processing plant or at the farm felt the biggest impact would be at the plant. Ten of eleven manufacturers who responded to a question concerning whether the biggest impact of membrane technology would be upon dairy products or dairy by-products (*e.g.*, on cheese as opposed to cheese whey) felt that the biggest impact would be in the by-product area.

THREE NEW PRODUCT CONCEPTS

Seven of ten manufacturers who responded to a question concerning whether membrane technology would be better at producing new dairy products or existing dairy products felt that the biggest impact would be upon new products.

The following new consumer product ideas were gleaned from the membrane industry:

- Protein-fortified, 2% reduced-fat milk can be made by a combination of whole milk, skim milk, and skim milk retentate. This product recipe had 18 per cent more protein than regular 2% reduced-fat milk without increased lactose levels.
- High-protein, low-lactose ice cream can be made by a combination of sweet cream, skim milk retentate, and nonfat dry milk. The desirability of this product results from substituting protein for lactose. The recipe evaluated had 48 per cent more protein and 32 per cent less lactose than regular ice cream.
- Nonfat yogurt can be made with more protein and therefore less stabilizers. This product can be made by a combination of skim milk, skim milk retentate, and nonfat dry milk. The particular product recipe evaluated had 15 per cent more protein and 21 per cent less lactose than regular nonfat yogurt.

These products all substitute protein for lactose. The addition of protein can bring more product body, better mouthfeel, and higher product viscosity. The reduction of lactose brings little in the way of reduced sweetness as lactose has only onesixth to one-third the sweetness of sucrose. Any such loss of sweetness can easily be countered by the addition of a small amount of sugar.

Focusing just on yogurt, the major benefits are two-fold. First, less product separation will occur. In other words, less liquid whey will form and separate from the yogurt curd. Second, if yogurt is made using non-dairy stabilizers, then product label-purity is compromised.

The non-dairy stabilizers which might be used for this purpose could include any of the following ingredients:

- Starch,
- Pectin,
- Gelatin,
- Vegetable gums (carboxymethyl cellulose,
- Locust bean, or guar), or seaweed gums (such as alginates or carrageenans).

To understand the important role of protein and why increased protein content is beneficial, consider the properties of two well-known dairy products, cheese and butter. The major difference between cheese and butter is that butter contains 80 per cent milkfat, while cheddar cheese contains approximately 32 per cent milkfat and 31 per cent protein. Even though butter contains less moisture than cheddar cheese, it remains a softer product. In addition, butter's weak texture makes it unsuitable for eating out of hand while a substantial amount of cheese is eaten in this fashion.

Finally, even though many different varieties of butter can be made, only one basic style is popular. In contrast, many different styles of hard cheese exist because protein is capable of conveying the tastes associated with different starter cultures and manufacturing methods. Can substituting protein for lactose reduce lactose levels enough to be beneficial to lactose intolerant consumers? The enzyme lactase is responsible for the digestion of lactose in the small intestine.

Individuals whose bodies produce insufficient lactase are said to be lactose intolerant. The severity of such lactose intolerance can vary from one individual to another.

For individuals with only mild intolerance, the reductions achieved by a proteinfor-lactose substitution could be beneficial. This would be particularly true for yogurt, which contains other beneficial bacteria to aid digestion (U.S. National Institutes of Health). To make each of these products, skim milk retentate (SMR) is needed.

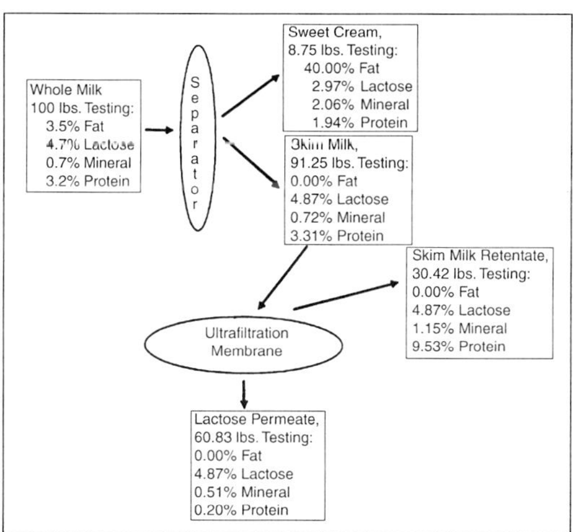

Fig. Flow and Mass Balance for the Manufacture of Skim Milk Retentate, the Building Block of New Dairy Products

SMR is produced by ultrafiltering skim milk to create a fluid isolate with a high protein-to-lactose ratio. As a result, SMR can reduce lactose content while increasing protein content. This means that protein can be substituted for lactose, improving the nutritional profile of dairy foods as well as their taste and texture. Further, it means that protein can substituted for the texture, functionality, and mouthfeel of butterfat.

COST OF CONCEPTS

U.S. dairy processing firms evaluate the purchase of new equipment very carefully due to budgetary constraints. Detailed system-cost information was provided by four equipment manufacturers and is shown in Table.

Table. Estimated Costs to Manufacture New Dairy Products

Characteristic	High-Protein 2% Butterfat Fluid Milk	High-Protein, Lower-Lactose Ice Cream Mix	High-Protein Nonfat Yogurt Mix
System Production/Day	375,000 Ibs. milk	200,000 lbs. mix	100,000 Ibs. mix
System Capital Cost[a]	$455,000.00	$1,240,000.00	$455,000.00
10-Yr. Depreciation (312 day basis)	$145.83/day	$397.44/day	$145.83/day
Operating Cost[a]	$675.00/day	$2,025.00/day	$675.00/day
Daily Capital and Operating Cost	$820.83/day	$2,422.44/day	$820.83/day
Capital and Operating Cost per Unit	$0.22/cwt.	$1.21/cwt.	$0.82/cwt.
Added Milk Cost[b]	$1.62/cwt.	$3.59/cwt.	$1.40/cwt.
Total Added Cost	$1.84/cwt. (or $0.16/gal.)	$4.80/cwt. (or $0.41/gal. Mix)	$2.22/cwt. ($0.19/gal. Mix)
Average Retail Price	$2.50/gal.	One gallon of mix will make four half- eightgallons ounce cups of selling for $3 00 each. each.	One gallon of mix will make 17 of ice cream yogurt selling for $0.50
Total Added Cost/ Average Retail Price	6.4%	3.4%	2.2%

Capital costs pertain to the membrane system and system hardware but exclude the cost for connection to utilities such as water, steam, and electricity. Specific capital costs include assembly, balance tanks, design engineering, electrical wiring, flow meters, gauges, installation, membrane housing, membranes, pipes, pressure gauges, process control computer, pumps, temperature recorders, and valves.

Capital costs were $455,000 for the fluid milk and yogurt membrane systems and $1,240,000 for the higher-capacity ice cream system. Membrane systems are relatively small and can usually be installed within an existing building; therefore, no cost for a building has been included. Capital costs were depreciated on a straight-line basis over ten years. Operating costs include those for membrane replacement, replacement of other parts, electricity, water, steam, sanitation materials, and labour. The third and final cost area pertains to the extra cost of the milk itself.

This results from inexpensive lactose being replaced by expensive protein. Note that although the ice cream system was more expensive, since it required greater capacity due to its greater substitution of protein for lactose, only one system-cost alternative was examined for each product. Debt was not included in the cost calculations.

The total added cost to make high-protein fluid milk was $0. 16 per gallon. Using an average retail price of $2.50 per gallon, the resulting cost increase relative to retail price is 6.4 per cent.

The added cost for the high-protein, lower-lactose ice cream mix was estimated at $0.41 per gallon of mix. Because of the incorporation of air, one gallon of ice cream mix will make four half-gallons of frozen ice cream. This equates to a cost increase of $0.1025 per halfgallon of frozen ice cream. Using an average retail price of $3.00 per half-gallon of ice cream, the resulting cost increase relative to retail price would be 3.4 per cent. The added cost for high-protein nonfat yogurt was estimated at $0.19 per gallon of mix. One gallon of yogurt mix will make 17 eightounce. cups of yogurt. Using an average retail price of $0.50 per cup, the resulting cost increase versus retail price would be 2.2 per cent.

SURVEY OF MILK PROCESSORS

In order to estimate the potential success of these new product concepts, a survey instrument was sent to U.S. dairy processors. Participants were informed of the particulars of the new product concept and supplied with the estimated cost of manufacturing the new dairy product. Background information was requested in a variety of areas including the respondent's opinion as to why customers purchased their existing dairy products, the importance of private-label products, the size of the firm, and the frequency of the respondent's contact with end-customers.

The survey also asked whether the firm presently employed any membrane technology for dairy purposes (only 10 per cent did so), whether the respondent thought consumers would buy the new product, and requested suggestions for increasing the probability of the new product's commercial success.

A total of 179 firms were contacted, of which 63 completed the survey for a total response rate of 35 per cent.

These 63 firms included 26 fluid milk processors, 21 ice cream manufacturers, and 16 yogurt manufacturers. The individuals surveyed were plant managers and/or those designated by each firm's receptionist as being most likely to make new product and/or new equipment decisions.

The survey instrument presented the new product idea, the equipment needed, the capital cost, the operating cost, and the increase in milk-component cost. Most of the interviews were initiated with a telephone call and then carried out by fax communication. Copies of the survey instruments are available from the authors upon request.

ANECDOTAL FINDINGS

Anecdotes reported by survey respondents are presented below. In all product areas, these comments illustrate a high level of cost and price sensitivity.

Based upon the results, one would expect this to be true for fluid milk. However, price sensitivity is also evident in the case of ice cream and even in the case of yogurt.

Comments received included:

- "Our current business isprice-driven, thus the supplier would have to absorb any cost increase." (From a bottler in the State of Washington.)
- "This product would have acceptance from only a small group of consumers." (From a bottler in Texas.)
- "We make a NFDM [Nonfat Dry Milk]-fortified product. At standard retail price it sells well. However, with a $0.05/gal. premium it does not sell well." (From a bottler in the State of New York.)
- "It takes a lot of consumer education and advertising to market a value-addedproduct at a premiumprice."(From a bottler in Kentucky.)
- "Adds cost. Our market area could not support this." (From an ice cream maker located in the North Central U.S.)
- "I am not sure customers would understand or care" (From an ice cream maker located in Wisconsin.)
- "This would take advertising." (From an ice cream maker in the High Plains.)
- "Customers currently love indulgence."(From an ice cream maker in California.)
- "Superiorflavour and texture at a competitive price will be a new product requirement." (From a yogurt maker in Ohio.)
- "I do not thinkyou can produce aproduct with enough improvement in taste/mouthfeel tojustify cost increase to the consumer." (From a yogurt maker in Michigan.)
- "Shelfplacement is critical." (From a yogurt maker in Illinois.)

STUDY IMPLICATIONS

These anecdotal findings reveal extreme concerns from dairy food manufacturers regarding highly elastic demand as well as the ability of consumers to perceive and/or pay for product differentiation. Because of cost, membrane technology investments for the purpose of producing a new product are likely to be scrutinized with great caution.

Among the three types of processors studied, statistical findings indicate that yogurt manufacturers are the most likely to consider a new functional-food formulation involving membrane technology. Concern about enhancing taste would likely be the motivation for making such a new product investment. This study has simply scratched the surface of an evolving industry process. Should membrane manufacturing technologies for new consumer dairy products become more widespread in the future, opportunities will then exist for economists to gain a greater understanding of this technology and its associated cost.

6

Animal Husbandry and Dairying

The contribution of animal husbandry and dairying to total gross domestic product (GDP) was 5.9 per cent in 2000-2001 at current prices. The value of output of livestock and fisheries sectors was estimated to be ₹1,70,205 crore during 2000-2001, which is 30.3 per cent of the total value of output of ₹5,61,717 crore from the agricultural and allied sectors. The contribution of the milk group alone (₹1,01,990 crore) was higher than wheat (₹47,091 crore) and sugarcane (₹27,647 crore).

It is estimated that almost 18 million people are employed in the livestock sector in principal (9.8 million) or subsidiary (8.6 million) status. Women constitute about 70 per cent of the labour force in livestock farming. The overall growth rate in the livestock sector is steady (around 4.5 per cent) in spite of fact that investment in this sector is not substantial. As the ownership of livestock is more evenly distributed with landless laborers and marginal farmers, the progress in this sector will result in a more balanced development of the rural economy.

REVIEW OF NINTH PLAN

CATTLE AND BUFFALO DEVELOPMENT

The broad frame-work of the cattle and buffalo breeding policy being followed since the mid-sixties envisaged selective breeding of indigenous breeds in their breeding tracts and use of such improved breeds for upgrading of the nondescript stock. While the States accepted the framework, appropriate implementation through field level programmes could not be done. Lack of interest in promoting Breed Organisation/Societies and related farmers' bodies contributed to the gradual deterioration of indigenous breeds.

Government intervention for breed improvement is not available to majority of owners of indigenous breeds of cattle. Eventually, the availability of good quality bulls needed for natural mating in breeding tracts became scarce, leading to further deterioration of indigenous breeds in these tracts. Production of quality indigenous bulls has been a long-neglected area and would require a major thrust in order to harvest the best male germplasm available in the country. The present production capacity of frozen semen doses is about 30 million against the estimated requirement of 65 million doses annually.

Except for a few pockets in important breeding tracts and in sperm stations, indigenous bulls of unknown pedigree and with poor quality semen are generally used. Crossbreeding, which was to be taken up in a restricted manner and in areas of low producing cattle, has now spread indiscriminately all over the country. Continuous emphasis on cross breeding with exotic breeds even in the tracts of indigenous breeds led to the nearextinction of some of the known breeds. Further, the indiscriminate use of contaminated semen or infected bulls results in the spread of sexually transmitted diseases like Infectious Bovine Rhinotracheitis (IBR) at an alarming rate.

MILK PRODUCTION

Milk production in India remained more or less stagnant from 1950 to 1970. Thereafter, it increased rapidly, reaching 84.6 million tonnes (mt) in 2001-02 (anticipated). But the Ninth Plan target of milk production (96.49 mt) was not achieved. The per capita availability of milk increased from 112 gm per day in 1973-74 to about 226 gm per day in 2001-02. However, it is still below the world average of 285 gm per day. Investment in the dairy sector in the Ninth Plan decreased significantly compared to the Eighth Plan.

Out of 168 Milk Unions, 58 Milk Unions (34.5 per cent) were running in loss as of March 2000. So far, the Government policy in the dairy sector has been to give preference to the establishment of milk processing plants linking rural milk producers to urban consumers through a network of cooperatives. Restrictions on establishing new milk processing capacity under Milk and Milk Products Order (MMPO) has now been removed. No policy measures have been undertaken so far to give a fillip to the unorganised sector involved in the production of Indian dairy products (like ghee, paneer, chhena, khoa etc.), which have tremendous potential in the export market in Asian and African countries.

EGG PRODUCTION

The Indian poultry industry has come a long way–from a backyard activity to an organised, scientific and vibrant industry. It is estimated that the egg production in the country is about 33. 6 billion numbers against the Ninth Plan target of 35 billion numbers. The most notable growth among the livestock products has been recorded in eggs and poultry meat. Since 1970-71, their output has grown at 5.87 per cent per annum.

The significant achievement in poultry development has come from the initiatives taken up by the private sector for commercial pure-line breeding. However, despite the huge investment made, mostly by the private sector, the poultry-processing sector is incurring losses. The status of the poultry sector as to; whether it falls under agriculture or industry, is somewhat ambiguous and, therefore, it has remained deprived of various benefits available to these sectors. Poultry farming should be declared as an agricultural activity.

The poultry production model in vogue (high input-high output using commercially developed strain of birds) has been primarily responsible for the rapid growth in production of eggs and broiler meat in the country, but it is successful mainly in large-scale units (more than 1,000 units of birds). Due to high feed cost, non-availability of credit and marketing support, most of the small farmers have become contract farmers and are exploited by middlemen.

Government intervention, by way of various support mechanisms, is now needed for the promotion of poultry in rural areas. Indigenous poultry breeds, including the improved strains that can survive with low quality raw feed and better resistance against diseases, can be reared under free range conditions by rural unemployed youth and women for some additional income and employment.

MEAT PRODUCTION

In India, meat production is largely a byproduct system of livestock production utilising spent animals at the end of their productive life. Meat production was estimated at 4.6 mt in 1998. Projects sanctioned during the Seventh and Eighth Plans for improvement/modernisation of abattoirs and carcass utilisation centres for fallen animals are still to be completed.

GOAT DEVELOPMENT

Despite the least attention from the planners, goat population in India has increased at the fastest rate among all major livestock species during last two decades. However, instead of increasing the goat population, emphasis should be laid on productivity per animal, organised marketing and prevention of emergence of new diseases like Peste des petits ruminants (PPR) which has led to higher mortality and abortion in goats. The goat improvement programme is to be given a push through extending credit to the poor landless farmers.

SHEEP PRODUCTION

During the last four decades, there has not been much increase in the sheep population. Production of wool has increased from 43.3 million kg in 1996-97 to 49.0 million kg (anticipated) in 2001-02. The Ninth Plan target of wood production (54.0 million kg) was not achieved. The fine wool production in the country is around 4 million kg against the demand of around 35 to 40 million kg. Indian wool is primarily used for the production of carpet, drugget, wall hangings etc. To enhance the quality and quantity of carpet wool, shepherds need incentives like credit, health coverage, breed improvement programmes and timely disposal of wool and surplus animals at a reasonable price.

PIG DEVELOPMENT

Pig husbandry is the most important activity in the animal husbandry sector in the northeastern region inhabited by tribal people. The region also has a substantial pig population, which constitutes around 25 per cent of the country's pig population. The bulk of the population is, however, of the indigenous type whose growth and productivity is very low. The major difficulty in pig development is the acute shortage of breeding males.

ANIMAL HEALTH

Since the Second Plan, efforts have been made to control diseases namely, Rinderpest, Foot and Mouth Disease, Haemorrhagic Septicemia, Black Quarter and Anthrax. Although Rinderpest has been eradicated from the country, the prevalence of the other diseases continues to be one of the major problems in the animal production programme.

Some of the emerging diseases like PPR, Bluetongue, Sheep Pox and Goat Pox, Classical Swine Fever, Contagious Bovine Pleuropneumonia, New Castle Disease (Ranikhet Disease) are causing substantial economic losses. The programme for creation of diseasefree zones was sanctioned in the Ninth Plan but was not implemented. The Department of Animal Husbandry and Dairying is also not well equipped with the necessary infrastructure and qualified technical manpower to execute the programmes and perform its mandatory duties and responsibilities like disease diagnosis and accreditation as per the international standards, development of an effective surveillance and monitoring system for diseases, mass immunisation against the most prevalent diseases etc.

Dovetailing the Animal Research Institutes of the Indian Council of Agricultural Research (ICAR) with the Department would not only improve its efficiency but also provide it with an effective delivery machinery to carry out its regulatory and certification authority functions, including the conservation of endangered breeds of livestock. The suggestion for the establishment of an Indian Council for Veterinary and Fisheries (ICVFR) by carving out the animal science and fishery institutes from ICAR has not yet materialised.

ANIMAL STATISTICS

The Livestock Census Scheme suffers from quantitative as well as qualitative problems. The present arrangements for conducting the Livestock Census in the States and Union Territories are not satisfactory in relation to timely collection of data and reporting. The Integrated Sample Survey Scheme for estimation of production of major livestock products also needs improvement.

CONSERVATION

The last few decades have witnessed serious erosion, and even extinction, of some indigenous animal breeds in the country. Many existing breeds are facing varying degrees of threat, endangerment and are heading towards eventual decimation. In all States, crossbreeding of cattle is now occupying a dominant position in the production programme and, in this process, the native cattle breeds, which are well adapted, have suffered wilful neglect resulting in their progressive elimination from the production system.

India is bestowed with rich domestic animal biodiversity, having 30 breeds of cattle, 12 breeds of buffalo, 20 breeds of goats, 40 breeds of sheep, eight breeds of camel, six breeds of horses, three breeds of pig and 18 of poultry. Besides, there are other species like equine, mithun, yak, turkey, ducks, etc. Indigenous breeds/types are rich in variability and are endowed with many positive traits like superior disease resistance, better tolerance to high heat and humidity and other characteristics suitable to particular agro-climatic environments. It has also been noted that indigenous breeds are more efficient in feed conversion particularly the crop residues and naturally available low quality roughages.

Indigenous breeds at risk are:

- Cattle:
  - Red Sindhi,
  - Sahiwal,
  - Tharparkar,
  - Punganur and
  - Vechur.
- Buffaloes:
  - Nili-Ravi,
  - Bhadawari and
  - Toda.
- Sheep:
  - Nilgiri,
  - Muzaffarnagri,
  - Malpura,
  - Chokla,
  - Jaisalmeri,
  - Munjal,
  - Changthangi,
  - Tibetan,
  - Bonpala from Sikkim and
  - Garrole sheep

- Goat:
  - Beetal,
  - Jamunapari,
  - Chegu,
  - Changthangi,
  - Surti and
  - Jakhrana.
- Camel:
  - Bacterian,
  - Jaisalmeri and Sindhi.
  - Yak Mithun
- Poultry:
  - All the 18 indigenous breeds of poultry are facing extinction.
  - The three important breeds are Aseel, Kadaknath and Naked Neck,

It has been globally recognised that conservation and improvement of native animal genetic resources are essential for sustainable development in agriculture and animal husbandry.

The conservation and improvement programme should be decentralised and each State/adjoining States where a breed exists should take necessary steps with the active involvement of institutions, Gaushalas, Non-government Organisations (NGOs) and Breed Societies. The efforts should, however, be effectively coordinated centrally.

Given the severity of the resource constraint, all the Central sector and Centrally sponsored schemes were subjected to zero-based budgeting during the Ninth Plan. The objective was to retain only those schemes that are demonstrably efficient and essential. The schemes that are similar in nature would be converged to eliminate duplication and resource flow would be linked to performance. Out of 41 schemes, 23 schemes were weeded out, one scheme was transferred and six schemes were merged.

Table. Average Annual Growth Rate of Milk and Egg Production 1950-51 to 2000-01

Year	Milk (%)	Eggs (%)
1950-51 to 1960-61	1.64	4.63
1960-61 to 1973-74	1.15	7.91
1973-74 to 1980-81	4.51	3.79
1980-81 to 1990-91	5.68	7.80
1990-91 to 2000-01	4.21	4.46

TENTH PLAN FOCUS AND STRAREGY

Animal husbandry and dairying will receive high priority in the efforts for generating wealth and employment, increasing the availability of animal protein in the food basket and for generating exportable surpluses.

The overall focus will be on four broad pillars viz:

- Removing policy distortions that is hindering the natural growth of livestock production;
- Building participatory institutions of collective action for small-scale farmers that allow them to get vertically integrated with livestock processors and input suppliers;
- Creating an environment in which farmers will increase investment in ways that will improve productivity in the livestock sector; and
- Promoting effective regulatory institutions to deal with the threat of environmental and health crises stemming from livestock.

The Tenth Plan target for milk production is set at 108.4 mt envisaging an annual growth rate of 6.0 per cent. Egg and wool production targets are set at 43.4 billion numbers and 63.7 million kg respectively. The allocation for animal husbandry, dairying and fishery is ₹2500 crore during the Tenth plan.

TRANSFER OF TECHNOLOGY

Use of technological and marketing interventions in the production, processing and distribution of livestock products will be the central theme of any future programme for livestock development. The generation and dissemination of appropriate technologies in the field of animal production as also health care to enhance production and productivity levels will be given greater attention.

Integration of Animal Research Institutes with the Department of Animal Husbandry and Dairying is essential to facilitate transfer of technology as well as to undertake sanitary and phyto-sanitary measures. This would provide an effective delivery machinery to the Department enabling it to work primarily as a regulatory body in the liberalised era.

HUMAN RESOURCE DEVELOPMENT AND EXTENSION

Sustainable rapid growth and development in this sector can only be ensured if the livestock owners, service providers, veterinarians and planners become knowledge based and acquire the ability to absorb, assimilate and adopt developments in the veterinary sciences and related technologies. Efforts will be made to improve the skills and competence of all stakeholders by involving village schools, veterinary colleges and universities in collaboration with the ICAR and its institutions including Krishi Vigyan Kendras (KVK), State Agricultural Universities and their field stations.

Steps will be taken to ensure that veterinary education is regulated as per the guidelines of the Veterinary Council of India. Introduction of animal science education (rearing of poultry, cattle, sheep, goat and pig) in the school curriculum will be one of the focus areas during the Tenth Plan. Training of para-veterinarians, Artificial Insemination (AI) technicians, laboratory technicians on a regular basis will be given priority.

Similarly livestock extension, which is primarily based on providing services and goods, will be treated differently from crop-related extension activities that are primarily based on transfer of knowledge. Livestock extension will be driven by technology transfer. As women play an important role in animal husbandry activities, deployment of women extension workers will be encouraged and they will work as links between farmers, the animal husbandry department and workers of NGOs.

INTEGRATION OF PROGRAMMES

Besides the Ministry of Agriculture, schemes relating to animal husbandry and dairying are being implemented by other ministries *viz.* Ministry of Rural Development, Ministry of Nonconventional Energy Sources etc. Many schemes operated by these ministries have similar and overlapping objectives and target the same population. Generic components like extension, training, and infrastructure get repeated in most of such schemes and are not complementary. Efforts will be made to consolidate and bring in convergence in these areas.

LIVESTOCK SERVICES

Most of the livestock services like artificial insemination/natural service, vaccination, deworming etc. are time-sensitive, which Government institutions, at times, are not able to deliver due to financial as well as bureaucratic constraints.

This necessitates the providing for efficient and effective decentralised services in tune with demands emanating from users. Efforts will be made to provide such services at the farmer's door, linked with cost recovery for economic viability. Availability of credit in time and technology support are the two important services needed for livestock development in the rural areas.

LIVESTOCK BREEDING STRATEGY

A national livestock breeding strategy needs to be evolved to meet the requirements of milk, meat, egg and other livestock products. Major thrust will be given to genetic upgradation of indigenous/native cattle and buffaloes using proven semen and high quality pedigreed bulls and by expanding the artificial insemination and natural service network to provide quality semen and other services at the farmer's level.

Improved bulls for natural breeding will be made available to private breeders, Gaushalas, NGOs and panchayats in remote and hilly areas. The programme of providing exotic males for improvement of sheep in the northern temperate region and pigs in the northeastern region will continue in the Tenth Plan. Financial and technological support would be needed to promote breeding programmes.

CONSERVATION OF BREEDS

Conservation of threatened breeds of livestock and improvement of breeds used for draught animals and packs would be one of the major goals of the Tenth Plan. It will be the national priority to maintain diversity of breeds and preserve those showing decline in numbers or facing extinction.

The improvement programme of indigenous breeds possessing desirable characteristics like disease resistance, heat tolerance, efficient utilisation of low quality feed etc. will be taken up. This is essential even for a sustainable crossbreeding programme. Steps will be taken to coordinate all the activities related to the efficient utilisation of draught animal power and animal by-products. Similarly efforts will be made to conserve indigenous birds and propagation of other birds like quail, guinea fowl and duck in those parts of the country where they are popular.

MILK PRODUCTION

The bacteriological quality of raw milk at the time of milking in India is comparable with that in the advanced dairying nations. Subsequently, however, the quality deteriorates due to improper handling of milk and lack of availability of infrastructure like all-weather roads, cooling facilities, potable water, regular electric supply and sewage disposal.

A holistic approach will be taken to address the issue of clean milk production, which is imperative for marketing and promoting export of dairy products. Steps will also be taken for development of unorganised milk sector that controls a significant portion of the liquid milk and sweetmeat market.

FODDER DEVELOPMENT

The importance of feed and fodder in livestock production hardly needs to be emphasised. Three major sources of fodder supply are crop residual, cultivated fodder and fodder from common property resources like forests, permanent pastures and grazing land. A significant portion of crop residue, particularly paddy and wheat straw, is being wasted. Emphasis will be given on enrichment of straw/stover, preparation of hay/silage to overcome fodder scarcities during the lean season, conversion of fodder into feed block to facilitate transport of fodder from surplus areas, establishment of fodder banks and promotion of chaff cutters.

The productivity as well as carrying capacity of public and forestland are decreasing due to improper management of common property resources and lack of coordination between the different agencies involved. For sustainable and economic livestock production, this problem will be addressed through scientific utilisation of traditional pastures and integration with the

Watershed Development Programme, especially for silvi-pastoral development.

For enhancement of grass production, measures will be taken to bring larger areas under joint forest management and treatment of wastelands and areas under problem soils. As the scope for increasing areas under cultivated fodder production is limited, efforts will be made to increase productivity through promotion of intensive fodder production technologies, quality fodder seed production by specialised agencies and use of wasteland for tree and bush based fodder production.

ANIMAL FEED

Oil cakes, maize and cereal by-products are important ingredients of animal feeds. Coarse grains and cottonseed are traditionally used as cattle feed. Measures will be taken to fill up the deficit in the requirement of feeds in quantitative and qualitative terms. At present, a very small portion of grains produced in the country is utilised for livestock and poultry feeding.

Rain-fed and arid zones present enormous prospects for production of feed grains. Steps will be taken to develop specifications for many agro by-products like mango seed kernel, mahowa cake, neem cake, soya pulp, whey powder etc. so that these could be utilised for feeding livestock. Quality control of animal feed will be given importance in the Tenth Plan.

ANIMAL HEALTH

Enhanced and sustainable productivity through improved animal health will be one of the major strategies during the Tenth Plan. After the successful eradication of Rinderpest disease, the major thrust will now be to adopt a National Immunisation Programme against the most prevalent animal diseases.

Animal disease diagnosis and accreditation as per the international standards, development of an effective surveillance and monitoring system for animal diseases, animal quarantine, certification and enforcement will be the major functions of the Department of Animal Husbandry and Dairying and necessary schemes will be evolved during the Tenth Plan. Further, measures will be taken to ensure that firms producing veterinary biologicals like vaccine, diagnostic kits etc. are following Good Manufacturing Practices (GMP) and meeting Good Laboratory Practices (GLP) requirements.

POULTRY PRODUCTION

The present system of production of commercial hybrid broilers and layers has become highly successful. To give a boost to export of poultry products, measures will be undertaken for the development of infrastructure like cold storage, pressured air cargo capacity and reference laboratory for certification of health and products. Programmes will be formulated to improve indigenous birds and promotion of backyard poultry farming which could help employment generation as well as economic empowerment of poor women in rural areas. There is tremendous scope for exporting poultry products produced from birds fed on organically produced feed.

CARCASS UTILISATION

Projects sanctioned during the Seventh and Eighth Plans for improvement/modernisation of abattoirs and carcass utilisation centres will be completed. Emphasis will be given on establishing/improving carcass utilisation centres for naturally fallen animals in rural areas.

MARKETING

The development of a marketing network and remunerative price support to the producers are great incentives for higher animal productivity and these will be encouraged for all types of livestock products. Even the advanced countries are giving direct and indirect price support to livestock farmers. Priority attention should also be given to improve processing, marketing and transport facilities for livestock products and value addition thereon. External markets are an extremely important source of demand and these will be tapped much more aggressively.

In order to encourage exports, licensing control for processing of livestock products/by-products will be repealed and restrictions on the export of livestock and its products will be removed. The immediate focus will be on export of animal and poultry products to Asian and African countries. The minimum requirements for sustainable export are creation of disease-free zones, organic farming and potable water. These will be made available in selected areas having large marketable surplus.

India has a large number of animal markets where livestock are traded but these are not developed on scientific lines. Market facilities are generally inadequate and, if available, are poorly maintained. Development of organised markets with adequate facilities will, therefore, be taken up. The concept of organic farming can also be extended to animal products.

Indian animals are reared in village pastureland and they are not generally treated with hormones, feed-antibiotics, or other drugs, so their products are healthy, wholesome and natural in every sense of the word. In rural India, cow dung and biomass are primarily used as manure. Initiative for export of 'Grassfed' animal products will be taken. Necessary infrastructure for certification procedures related to organic animal farming will be promoted.

QUALITY AND SAFETY OF LIVESTOCK PRODUCTS

Quality and safety of livestock products depend upon a quality and safety assurance system for which legislation for setting up standards, corresponding to Codex standards, is obligatory. These do not exist nor is there any method for reviewing and rationalising the quality and safety guidelines. Efforts will also be made for harmonisation of infrastructure facilities for testing food quality and safety with international standards.

DATABASE

Currently, there is absence of a lot of data like those relating to breed-wise milk production of cattle and buffalo, egg production from commercial farms and households, cost of production of milk, egg and wool, availability of livestock resources etc. A National Animal Health and Production Information System will be established with the active involvement of research Institutions, Government departments, panchayati raj institutions (PRIs), urban local bodies (ULBs), private industries, cooperatives and NGOs. This will work as the national database.

ANIMAL WELFARE

Animal welfare is also related directly with the productivity of animals. The well-being of animals is affected during management under the intensive production system, in the animal market, during handling and transportation, rearing of buffalo male calves in urban areas etc. There is a great deal of wastage, as well as animal suffering due to ill-designed agri-implements, carts and implements attached to animals.

Efforts will be made to strengthen the institutions working on a livestock care system so that they can ensure and promote animal care and well-being. Research and technology development will be taken up for enhancing efficiency and reducing drudgery of animals by improving the design of carts, yokes, implements and toolbars used in agriculture. A good example is the buffalo-drawn bogey fitted with rubber tyre and bearings.

DEVELOPMENT OF LOCATION SPECIFIC ANIMALS

Camel will continue to be important in desert areas for quite some time. Effective support for providing nutrition and health cover is needed for its improvement. The Department of Animal Husbandry will continue its programme for improvement of better studs both for horses and donkeys used for transport in hilly areas.

Horse riding is now becoming an integral part of amusement parks and this will be encouraged as a niche industry. To encourage the breeding of horses, mules and asses, technological and financial support will be extended to entrepreneurs. Animals indigenous to specific agro-climate regions like Yak and Mithun will be developed.

CAPITAL FORMATION

Public sector lending in the livestock sector is low and inadequate credit support leads to poor capital formation. As the organised financial sector is unwilling to finance livestock programmes that are not in their interest, especially after the initiation of financial sector reforms, the livestock farmers are mainly dependent on the financial intermediaries and they end up bearing a higher interest rate than would be available otherwise.

Attempts would be made to create a favourable economic environment for increasing capital formation and private investment. Financial institutions would actively participate in livestock credit programmes through standardised ready-made bankable projects with back-ended subsidy. Creation of a venture capital fund is needed to assist the private entrepreneur in establishing units that could provide services and goods at the district/ block level.

THE PATH AHEAD

The programmes that will be emphasized during the Tenth Plan are:

- The major thrust will be on genetic upgradation of indigenous/native cattle and buffaloes using proven semen and high quality pedigreed bulls and by expanding artificial insemination and natural service network to provide services at the farmer's level. Production of progeny-tested bulls in collaboration with military dairy farms, government/institution farms and gaushalas will be taken up.
- Conservation of livestock should be the national priority to maintain diversity of breeds and preserve those showing decline in numbers or facing extinction.
- After the successful eradication of Rinderpest disease, the focus would now be to adopt a national immunisation programme to control prevalent animal diseases. Efforts will be made for the creation of disease-free zones.
- Development of fodder through cultivation of fodder crops and fodder trees, regeneration of grazing lands and proper management of common property resources.
- Improvements of small ruminants (sheep and goat) and pack animals (equine and camel) should be taken up in the regions where such animals are predominant.
- Building infrastructure for animal husbandry extension network. Panchayats, cooperatives and NGOs should play a leading role in generating a dedicated band of service providers at the farmer's doorstep in their respective areas
- Strengthening infrastructure and programmes for quality and clean milk production and processing for value addition.
- Programmes would be implemented to improve indigenous birds and promotion of backyard poultry in rural areas.
- An information network would be created based on animal

production and health with the active involvement of Research Institutions, Government departments, private industries, cooperative, and NGOs.

- Strengthening of veterinary colleges as per the norms of Veterinary Council of India. Strengthening of Department of Animal Husbandry and Dairying is also crucial if it has to work as a regulatory and monitoring authority.
- A regular interaction between the Department of Animal Husbandry and Dairying and research institutes like the Indian Veterinary Research Institute, National Dairy Research Institute, Institutes on cattle, buffalo, sheep, goat, equine and camel.

FISHERIES

The fisheries sector is one of the important sectors in the socio-economic development of the country. More than six million fishermen and fish farmers, a majority of whom live in 3937 coastal villages, besides fishermen hamlets along major river basins and reservoirs in the country, depend on fisheries and aquaculture for their livelihood. The sector has also been one of the major contributors to foreign exchange earnings through exports. India is the third largest fish producer in the world and second in inland fish production.

The fisheries sector contributes ₹19,555 crores to national income which is 1.4% of the total GDP. The country is endowed with an Exclusive Economic Zone (EEZ) extending to 20.2 lakh sq. kms. with a continental shelf area of about 5.2 lakh sq. kms. having about 8118 kms. coastal length with some of the richest fishing grounds in the world. The estimated potential for fish production from inland water bodies is about 4.5 million tonnes(mt). The main inland fishery resources include about 1.20 million hectares (m ha.) of brackish water area, about 23.81 lakh ha. of fresh water ponds and tanks, about 7.98 lakh ha. lakes and about 20.31 lakh ha. of reservoirs, besides about 1,91,000 kms of rivers and canals.

REVIEW OF THE NINTH PLAN

During the last five decades, fish production has increased with an annual growth rate of 4.1 per cent. Fish production touched 5.67 mt in 1999-2000 and is estimated to be about 5.66 mt in 2000-01. It is likely to reach a level of 6.12 mt by the end of the Ninth Plan, which is much below the target of 7.04 mt. This is because of slow progress in the fish production to the extent of 1.44 per cent per annum during the first four years of the Ninth Plan.

At present, resource-wise (reservoirs/rivers/ponds/tanks etc.) data on fish production and productivity are not available in the country. In the absence of any major initiative for strengthening of infrastructure, fish seed production remained almost static (16,000 million fry per annum) during the first four years of the Ninth Plan.

INLAND FISH PRODUCTION

The share of inland fishery sector in fish production, which was 29 per cent in 1950-51 (0.22 mt), has increased to about 50 per cent in 1999-2000 (2.84 mt). In spite of this, the present level of fish production in the country is about 67 per cent of the estimated potential of 8.4 mt. There is enormous scope both for augmentation of production potential as well as enhancement of productivity in the inland fishery sector. The 429 Fish Farmers Development Agencies (FFDAs) have covered about 5.67 lakh ha. (inclusive of 1.70 lakh ha. in Ninth Plan) of the total water area under scientific fish culture and trained 6.51 lakh fish farmers (1.11 lakh in Ninth Plan).

But the average productivity from waters covered under this programme remained almost static at about 2.2 tonnes/ha./year during the Ninth Plan period. States like Andhra Pradesh, Punjab and West Bengal have shown better response and faster development. The highest productivity of about 5 tonnes/ha/annum from FFDA ponds/tanks has been achieved in Punjab. About 6240 ha. was brought under brackish water aquaculture activities during the Ninth Plan through 39 Brackish Water Fish Farmers Development Agencies (BFDAs). The performance of the programme has also been affected due to litigation.

MARINE FISH PRODUCTION

Marine capture fisheries play a vital role in India's economy. The sector provides employment and income to nearly two million people. Marine fish production level has risen from 0.53 mt in 1950-51 to 2.81 mt in 2000-01 with a growth rate of 3.43 per cent. Most of the major commercially exploited stocks are showing signs of over exploitation. Problems of juvenile finfish mortality and bycatch discards increased with the intensification of shrimp trawling. Plateauing of catches and over-fishing at several centers and inter sectoral conflicts in the coastal belts have highlighted the need for caution.

Proper management of coastal fishery resources with suitable enforcement mechanisms like uniform ban on fishing during monsoon which is considered the breeding season for majority of commercial species, regulation on craft and gears etc. are the priority issues in the sector to allow for its rational exploitation. The development of the deep-sea fishery industry is of concern to the entire marine fishery sector because it would have considerable impact on the management of near-shore fisheries, shore-based infrastructure utilization and post-harvest activities both for the domestic market and exports. With the growing demand for sea food, it becomes imperative that the current level of marine fish production from the exploited zone to be sustained by closely monitoring the landing and the fishing effort and by strictly implementing the scientific management measures.

INFRASTRUCTURE

The existing fishing harbours and infrastructures need to be modernized to meet minimum international standards necessary for fish quality assurance. Under the Fisheries Extension and Training Programme 28 training centers and 15 awareness centers have been established for the benefit of fishermen and fish farmers during the Ninth Plan.

Research projects in the area of aquaculture and marine biotechnology are supported to strengthen the gap in the areas of fish health and disease diagnostics, transgenic aspects, cell and tissue culture, intensive prawn culture, carp-culture, feed and seed production, bio-active compounds and development of culture technology in non-conventional species etc. by the Department of Bio-technology during the Ninth Plan.

TENTH PLAN FOCUS AND STRATEGIES

DEVELOPMENT OF FISHERIES

The major thrust during the Tenth Plan will be on integrated development of riverine fisheries, habitat restoration and fisheries development of upland waters, development of reservoir fisheries, management of coastal fisheries, deep-sea fisheries with equity participation, vertical and horizontal development of aquaculture productivity, infrastructure development and improved post-harvest management, policy intervention including monitoring, control and surveillance. The Tenth Plan has proposed a fish production target of 8.19 mt envisaging a growth rate of 5.44 per cent per annum (marine 2.5 per cent and inland 8.0 per cent).

DEVELOPMENT OF AQUACULTURE

In the recent years, there has been a spurt in the growth of aquaculture in the country. The inland fisheries sector has registered an impressive growth rate of 6.55 per cent per annum in the 1990s. However, in spite of the vast resources of culturable water bodies as well as availability of proven technology for aquaculture, the levels of production and productivity are not adequate and there is a large gap between the potential and actual yields. Therefore, increase in productivity and production of fish/shrimps from freshwater and brackish water areas under ongoing programmes would continue during the Tenth Plan.

The present production level of about 2.2 tonnes/ha./year from fish farming will be raised considerably by adopting existing advance technology. Programmes will be devised to develop fisheries in fallow derelict water bodies, waterlogged areas, saline waters, lakes, beels, etc. for enhancing fish production. Aquaculture activities will also be taken up for development of cold-water fisheries in the hill areas of the ecologically fragile zone.

On the basis of experience of pilot projects taken up for fisheries development in reservoirs during the terminal year of the Ninth Plan, programme to enhance fish production will be formulated on a large scale during the Tenth Plan. An integrated approach to marine and inland fisheries, designed to rational exploitation and to promote sustainable aquaculture practices, will be adopted. Bio-technological applications in the field of genetics and breeding, hormonal application, immunology and disease control will receive particular attention for increased aquaculture production.

SEED AND FEED DEVELOPMENT

Seed and feed are critical inputs required for the development of fisheries and aquaculture for enhancing production and productivity. Research and development (R&D) programmes will be taken up for production of quality fish/shrimp seed and feed. The present level of fish seed production of 16,000 million fry will be raised to 25,000 million fry by the end of the Tenth Plan at an 8 per cent growth rate per annum. Diseases-free and diseases-resistant fish/shrimp seed will be ensured with strict quarantine measures.

Besides, adequate infrastructure will be required for increasing production and productivity of other commercially important fishes/prawn such as freshwater prawn, catfish, sea bass, grey mullet, grouper, snapper, chanos, etc. for diversifying fishing activities during the Tenth Plan. The Research Institutes under the ICAR like Central Institute of Fisheries Education (CIFE), Mumbai, Central Marine Fisheries Research Institute (CMFRI), Kochi, and Central Institute of Fresh Water Aquaculture (CIFA), Bhubaneswar, have developed technology for pearl culture, which needs to be taken up on a commercial basis through concerted efforts for further development during the ensuing Plan period.

TRAINING OF FISHERWOMEN

Traditionally, women have played an important role in the fishery sector, and they have a much larger role to play in the emerging scenario of fisheries and aquaculture development. One of the important ways to improve the status of fisher-women in a community is to train them to improve their participation in their own development. Programmes for human resource development with emphasis on training and skill development in post-harvest/processing and marketing activities particularly for fisherwomen besides other income generating revenues will be taken up. Emphasis will be laid on the development of marketing infrastructure and techniques of preservation/storage and transportation with a view to reducing post-harvest losses and ensuring a better return to the grower.

STRENGTHENING OF DATABASE

Notwithstanding the existing efforts made by several agencies, the fisheries database is poor and needs considerable strengthening. In the inland sector, the priorities are standardization of methodologies for estimation of catch from the diverse aquatic resources and establishing mechanisms for regular collection and dissemination of data by States and Union Territories.

In the marine sector, the existing methodo-logies need revision and also subsequent re-orientation of the Departments of Fisheries on collection and estimation of methodologies. To strengthen the efforts in this direction, the use of remote sensing and Geographical Information System (GIS) in estimation of resource size and productivity also needs to be integrated in the existing programmes of fisheries catch statistics.

OVEREXPLOITATION OF COASTAL RESOURCES

A major emphasis will be placed on positive and purposeful checks on over exploitation of resources in the near shore areas through appropriate regulations on the number of fishing vessels, their operational areas, ban on monsoon fishing/close season, mesh size, use of the right type of fishing gear and other such restrictions to prevent uneconomic and oversize fishing.

EXCLUSIVE ECONOMIC ZONE

Exploitation of offshore resources in the EEZ will be considered in terms of both the resource available and the infrastructure. Along with the absolute right on the EEZ, India has also acquired the responsibility to conserve, develop and optimally exploit the marine living resources within this area. Efforts will be made to exploit fishery resources in the EEZ on a priority basis. Satellite-assisted Vessel Monitoring System (VMS) will be helpful in the EEZ for both Indian and foreign fishing vessels. This would ensure the safety of fishers and vessels, and also provide emergency help whenever required.

This would also help in the collection of fishery-related technical data as well as determining the number of fishing vessels required in a particular area for exploiting the available fishery resources. Formulation and introduction of a new deep sea fishing policy consistent with the national interest to exploit fishery resources in the EEZ should be given top priority. The present gap in the potential and current exploitation has several repercussions, the more important of which is leaving the EEZ opening to other neighbouring countries like Nepal, Bhutan etc. and owners of foreign fishing vessels which may take advantage of the situation.

Besides, even land locked neighbouring countries like Bhutan, Nepal etc. may stake their claim legally unless we put our efforts together on under-exploited marine resources in the Indian EEZ. Efforts are also needed to maintain World Trade Organisation (WTO) catch levels by rational exploitation of our resources and to counter measures taken by neighbouring countries like Pakistan in collaboration with USA which is resulting in the over-exploitation of resources in the adjoining areas and there by curtailing our rights in these areas. Besides it should also be ensured that suitable measures are taken to exploit resources beyond the EEZ so that we put our due stake in the international waters alongwith other countries.

INVESTMENT

Increasing public/private investment is needed for strengthening infrastructure for diversifying fisheries and aquaculture activities enhancing fish production and productivity. Enhanced public investment is also required in research programmes, strengthening infrastructures for training, post-harvest, marketing etc. Setting up of minor fishing harbours and creation of common facilities for maintenance and usage of dredgers by the Government should be given priority for improvement of infrastructure facilities in the marine fishery sector.

Product development by value addition of low quality fish and development of products like chitosene out of wastes like prawn shells, products out of fish bladder etc. need to be encouraged. Private sector investment in fisheries will also be encouraged particularly in seed and feed production, adopting existing technologies for higher production, human resource development, post-harvest management and marketing. For sustainable development of coastal areas, establishment of agro-aqua farms along coastal regions, linking ecological security with livelihood security would be encouraged by States/NGOs.

Such farms involve concurrent attention to culture and capture fishery and forestry and agroforestry programmes. Besides, conservation of fisheries resources, these farms would also be used for demonstrations of diversifying activities of different techniques to be used for fishing operations. Emphasis would be given for technological upgradation of the traditional fishing sector with improved motorised crafts and gears for the development of coastal fisheries and for the introduction of new generation of fishing vessels, for development of off-shore fishing with modern communication equipments to ensure safety of fishermen while out at sea etc.

Proper credit and technological support for standard bankable projects and ventures by small fishermen groups in the inland sector and setting up of cooperative marketing network in marine sector should be ensured through institutional finance from the National Bank for Agriculture and Rural Development (NABARD) and National Cooperative Development Corporation (NCDC).

NEW INITIATIVES

The new initiatives for development of fisheries during the Tenth Plan would be to increase production and productivity from deep seas, inland capture fishery resources like rivers, canals etc. and from culture sources like reservoirs, beels, ox-bow lakes, measures for replenishment of fishery resources through mariculture etc. Besides, development of infrastructural facilities for better post-harvest management, technology for sustainable aquaculture, setting up of cold storage and marketing network through viable fishermen cooperatives etc., are also proposed to be taken up to ensure better livelihood for fishers and enhance export promotion for economic development of the country.

THE PATH AHEAD

The main thrust for fisheries development during the Tenth Plan would be to utilise the full potential of inland fishery resources as well as deep seas to increase per capita consumption to a substantial level from the present level of 9 kg. per head per annum.

Special emphasis will be given on:

- Increasing the depth of fishing harbours especially for small fishermen using dredgers and the upgradation of hygienic conditions there.
- Strengthening of data base and information networking in the fisheries sector for standardisation of methodologies and estimation of catch from diverse aquatic resources.
- Aquaculture and development of capture fisheries of inland water resources.
- Measures will be taken to increase fish production from the deep sea marine sector.
- Infrastructure development, post harvest management for marketing by setting up of model fish markets and establishment of cold chain through viable fishermen cooperatives.
- Popularisation of pearls developed by CIFA, CMFRI etc. and value added products developed by the Central Institute of Fisheries Technology (CIFT), Kochi and Integrated Fisheries Project (IFP), Kochi made out of low value fish with suitable credit/subsidy support.
- Welfare measures for fishers will be strengthened to ensure their safety at sea etc. and also to involve more women in fisheries sector.
- Research and technology needs in fisheries institutes to be upgraded to meet the growing demands.
- Formulation of a comprehensive deep sea fishing policy and passing of the Aquaculture Authority Bill in Parliament to be expedited for rational exploitation of deep sea fishery resources and sustainable aquaculture development.
- Strategy for an effective enforcement mechanism is needed to prevent poaching in the EEZ and thereby safeguard our resources.
- Suitable mariculture programmes need to be undertaken for commercially important fin/shell fish species for replenishment of resources in our seas.
- Setting up of disease control laboratories and quality certification centres to ensure international standards for fishery products.
- Technologically improved fishing boats with proper communication network etc. to be introduced for the benefit of small fishermen.

7

Husbandry, Housing, and Biosecurity

Proper management is essential for the well-being of the animals, the validity and effectiveness of research and teaching activities, and the health and safety of animal care personnel. Sound animal husbandry programmes provide systems of care that permit the animals to grow, mature, reproduce, and be healthy. Specific operating procedures depend on many factors that are unique to individual institutions. Well-trained and properly motivated personnel can often achieve high quality animal care with less than ideal physical plants and equipment.

FACILITIES AND ENVIRONMENT

ENVIRONMENTAL REQUIREMENTS AND STRESS

Domestic animals are relatively adaptable to a wide range of environments. Domestication is a continuing process. Genetic strains of animals selected for growth or reproduction in different environments under varying degrees of control are used currently for much of the production of livestock and poultry. These strains of animals are sometimes very different from the breeds or strains from which they were originally derived. Agricultural animals may be kept in extensive environments where they reside in large areas outdoors.

They may also be kept in intensive environments where they are confined to an area that would not sustain them were the environment not controlled and where food, water, and other needs must be provided to them. Individual animals may be moved during their lives from extensive to intensive systems or *vice-versa*. Species requirements for domesticated animals are thus variable and depend both on the genetic background of the animals and their prior experience.

CRITERIA OF WELL-BEING

Various criteria have been proposed to identify inappropriate management and housing conditions for agricultural animals. For example, in poultry, significant feather loss that is not associated with natural mating or natural molting is widely accepted as an indication that birds are experiencing stressful conditions. More sophisticated measures of stress are not necessarily superior and may even yield confusing results and lead to inaccurate conclusions.

For instance, plasma corticosteroid concentrations of hens residing in spacious floor pens may be similar to those in high-density cages, even though other criteria may indicate that the caged hens are adversely affected by their environment. During stressful social situations, resistance to virus-induced diseases may be depressed, but resistance to bacterial infections and parasites may be increased. Some researchers have placed emphasis on behavioural criteria of well-being, although others have pointed out the difficulties of interpretation involved.

In the same way, some researchers have suggested that depressed performance of individuals, independent of economic considerations, is a relatively sensitive reflector of chronic stressors, but Hill was less convinced using the same parameters. Animal well-being has both physical and psychological components. No single objective measurement exists that can be used to evaluate the level of well-being associated with a particular system of agricultural animal production. There is consensus, however, that multiple integrated indicators provide the best means of assessing well-being.

Indicators in 4 categories ries are generally advocated:

1. Behaviour patterns;
2. Pathological and immunological traits;
3. Physiological and biochemical characteristics; and
4. Reproductive and productive performance of the individual animal.

A judgment as to the balance of evidence provided by these indicators has been used, when available, as the basis for the recommendations in this guide. D.C. Hardwick postulated and Duncan developed the idea that an acceptable level of animal welfare exists over a range of conditions provided by a variety of agricultural production systems, not under just one ideal set of circumstances.

Improvements in certain environments may increase animal well-being somewhat, but any point in the range would still be considered acceptable with respect to animal welfare. Good management and a high standard of stockmanship are important in determining the acceptability of a particular production system and should be emphasized in agricultural animal research and teaching facilities.

MACROENVIRONMENT AND MICROENVIRONMENT

Animal well-being is a function of many environmental variables, including physical surroundings, nutritional intake, and social and biological interactions. Environmental conditions should be such that stress, illness, mortality, injury, and behavioural problems are minimized. Particular components of the environment that need to be taken into account include temperature, humidity, light, air quality, space, social interactions, microbe concentrations, noise, vermin and predators, nutritional factors, and water.

Environmental Enrichment for further information Physical conditions in the room, house, barn, or outside environment constitute the macroenvironment; the micro-environment includes the immediate physical and biological surroundings. Different microenvironments may exist within the same macroenvironment. Both microenvironment and macroenvironment should be appropriate for the genetic background and age of the animals and the purpose for which they are being used. Domestic animals readily adapt to a wide range of environments, but some genetic strains have specific needs of which the scientist should be aware and for which accommodation should be made. Even in relatively moderate climatic regions, weather events such as floods, winter storms, and summer heat waves may require that animals have access to shelter. If trees or geographic features do not provide enough protection, artificial shelters and windbreaks or sunshades should be provided.

GENETIC DIFFERENCES

Some strains of agricultural animals may have requirements that differ substantially from those of other stocks of the same species. Some strains of pigs, for example, are particularly susceptible to stress because they carry a gene that causes malignant hyperthermia when they experience even mild stress. Transgenic animals may also have special needs for husbandry and care. Practices to ensure the well-being of special strains should be established independently of those made for the species in general.

SPACE REQUIREMENTS

Floor area is only one of the components that determine the space requirements of an animal. Enclosure shape, floor type, ceiling height, location and dimensions of feeders and waterers, features inside the enclosure, and other physical and social elements affect the amount of space sensed, perceived, and used by the animals in intensive management systems.

When possible, animals in stanchions, cages, crates, or stalls should be allowed to view one another, animal care personnel, and other activities where this would not interfere with research or teaching objectives. Determination of area requirements for domestic animals should be based on body size, head height, stage of life cycle, behaviour, health, and weather conditions.

All area recommendations in this guide refer to the animal zone. Unless experimental or welfare considerations dictate otherwise, space should be sufficient for normal postural adjustments, including standing, lying, resting, self-grooming, eating, drinking, and eliminating feces and urine. When animals are crowded, body weight gain and other performance traits may be depressed and the animals may show altered levels of aggressive behaviour.

TEMPERATURE, WATER VAPOUR PRESSURE, AND VENTILATION

Air temperature, water vapour pressure, and air velocity are some of the most important factors in the physical environment of agricultural animals. In addition, factors related to animal health and genetics affect the thermal balance of animals and thus their behaviour, metabolism, and performance. Most agricultural animals are quite adaptable to the wide range of thermal environments that are typically found in the natural outdoor surroundings of various climatic regions of the continental United States.

The range of environmental temperatures over which animals use the minimum amount of metabolizable dietary energy to control body temperature is termed the thermoneutral zone. Homeothermic metabolic responses are not needed within this zone. Temperature and vapour pressure ranges vary widely among geographic locations. The long-term well-being of an animal is not necessarily compromised each time it experiences cold or heat stress.

However, the overall efficiency of metabolizable energy use for productive purposes is generally lower outside the thermoneutral zone than it is within the zone. The preferred thermal conditions for agricultural animals lie within the range of nominal performance losses. Actual effective environmental temperature may be temporarily cooler or warmer than the preferred temperature without compromising either the overall well-being or the productive efficiency of the animals.

Evaluation of thermoregulation or of heat production, dissipation, and storage can serve as an indicator of well-being in relation to thermal environments. The thermal environment that animals actually experience represents the combined effects of several variables, including air temperature, vapour pressure, air speed, surrounding surface temperatures, insulative effects of the surroundings, and the age, sex, weight, infectious status, transgenic modification status, adaptation status, activity level, posture, stage of production, body condition, and dietary regimen of the animal.

To overcome shortcomings of using ambient temperature as the only indicator of animal comfort, thermal indices have been developed to better characterize the influence of multiple environmental variables on the animal. The temperature-humidity index (THI), first proposed by Thom has been extensively applied for moderate to hot conditions, even with recognized limitations related to airspeed and radiation heat loads.

At the present time, the THI has become the de facto standard for classifying thermal environments in many animal studies and selection of management practices during seasons other than winter. The THI has further been used as the basis for the Livestock Weather Safety Index to describe categories of heat stress associated with hotweather conditions for livestock exposed to extreme conditions. Categories in the LWSI are alert, danger, and emergency. Additionally, THI between 70 and 74 is an indication to producers that they need to be aware that the potential for heat stress in livestock exists. The index {wind chill temperature index (°C) = 13.12 + (0.6215 × AT)-[11.37 × (WSPD)0.16] + [0.3965 × AT × (WSPD)0.16]}, where AT = air temperature, °C, and WSPD = wind speed, m/s, is a physiological based model and accounts for inherent errors in the earlier wind chill index (WCI), which was not based on heat transfer properties of body tissues.

However, the old WCI closely mimicked heat loss and equivalent temperature equations reported by Ames and Insley for sheep and cattle. Equations developed by Ames and Insley accounted for heat transfer through pelts and hides sections of previously harvested animals; however, they did not account for fat cover and other regulatory processes utilized in mitigating cold stress. In addition, body heat loss due to wind will be proportional to the surface area exposed and not the entire surface area of the body.

This error was also inherent in the old WCI. A ventilation system removes heat, water vapour, and air pollutants from an enclosed animal facility at the same time that it introduces fresh air. Adequate ventilation is a major consideration in prevention of respiratory and other diseases. Where temperature control is critical, cooling or heating may be required to supplement the ventilation system. For certain research projects, filtration or air conditioning may be needed as well.

Typically, ventilation is the primary means of maintaining the desired air temperature and water vapour pressure conditions in the animal microenvironment. The amount of ventilation needed depends on the size, number, type, age, and dietary regimen of the animals, the waste management system, and atmospheric conditions. Equipment and husbandry practices that affect heat and water vapour loads inside the animal house also should be considered in the design and operation of the ventilation system.

Ventilation rates in enclosed facilities should increase from a cold-season minimum to a hot-season maximum. It is important to recognize the approximately 10-fold increase in ventilation rate from winter to summer that is required in a typical livestock or poultry house. Because the animals themselves are the major source of water vapour, heat, and odorous matter, ventilation rate calculated on the basis of animal mass is more accurate than that based on airexchange rate guidelines. Relative humidity is ordinarily the parameter used to manage the air moisture content.

Hot weather ventilation rates should be sufficiently high to maintain the relative humidity below 80% in an enclosed animal house except for situations in which high relative humidity does not cause animal health concerns. Conversely, ventilation rate during cold weather should be sufficiently low to ensure that the relative humidity does not fall to a level that causes animal health concerns, unless needs for air quality or condensation control necessitate a higher rate. Atmospheric humidity does not ordinarily become a significant factor in effective environmental temperature until the air temperature approaches the temperature of the animal's surface, in which case the animal will depend almost entirely on evaporative heat loss to maintain thermal equilibrium with the environment. The use of fans to promote air movement can be beneficial during hot weather if there is too little natural air movement.

Direct wetting is effective in decreasing heat stress on cattle and pigs; however, it can cause the death of poultry. Wetting is best accomplished by water sprinkled or dripped directly on the animals. Misters and evaporative coolers specifically designed to reduce air dry-bulb temperature are also used to reduce heat stress on agricultural animals. Correctly designed and maintained sunshades protect animals from heat stress by reducing solar radiation load. Trees, if available, are ideal sunshades. Artificial, roofed shades are acceptable.

Mechanical ventilation requires proper design and operation of both air inlets and fans for proper distribution and mixing of the air and thus for creating uniform conditions throughout the animal living space. Mechanical ventilation, with fans creating static pressure differences between inside and outside the house, brings in fresh air and exhausts air that has picked up heat, water vapour, and air pollutants while passing through the building.

Mechanical ventilation, if properly designed, provides better control of air exchange for enclosed, insulated animal houses in colder climates than does natural ventilation. The effectiveness of natural ventilation in cold climates will depend on the design and orientation of the enclosure, as well as the species and number of animals housed and the stage of their life cycle. Natural ventilation uses thermal buoyancy and wind currents to vent air through openings in outside walls or at the ridge of the building.

Natural ventilation is especially effective for cold animal houses in moderate climates; however, insulated walls, ceilings, and floors are often recommended to minimize condensation. The air exchange rate needed to remove the water vapour generated by animals and evaporation of water from environmental surfaces often brings air temperature inside such houses down to values near those outdoors.

If waterers and water pipes are protected from freezing, the practical low operating temperature is the point at which manure freezes, although this temperature would be too cold for some species or stages of the life cycle. Automatic curtains or vent panels, insulated ceilings, and circulating fans help to regulate and enhance natural ventilation systems.

During cold weather, ventilation in houses for neonatal animals should maintain acceptable air quality in terms of water vapour and other pollutants without chilling the animals. Air speed should be less than 0.25 m/s past very young animals. There should be no drafts on young poultry or pigs. During hot, warm, or cool atmospheric conditions, ventilation of animal houses should maintain the thermal comfort of the animal to the extent possible. Ideally, the ventilation rate should be high enough to prevent indoor temperature from exceeding outdoor temperature by more than 3°C when the atmospheric temperature is above 32°C for small animals and above 25°C for larger ones. In arid and semi-arid regions where the potential for evaporative heat loss is great, air temperature may peak at over 43°C for 1 or 2 d or longer without affecting animal well-being if animals have been acclimatized by chronic exposure.

Ventilation system design should be based on building construction and the rates of water vapour and heat production of the animals housed. The frame of reference is the animal microenvironment. For example, the outdoor calf hutch is a popular accommodation for dairy replacement heifer calves in most parts of the continental United States. Although the hutch provides a cold microenvironment for calves during winter in northern latitudes, the calf is nonetheless comfortable if cared for correctly.

In closed houses during hot periods, additional ventilation capacity may be necessary. In enclosed animal houses, both environmental temperature and air quality depend on the continuous functioning of the ventilation system. An automatic warning system is desirable to alert animal care and security personnel to power failures and out-of-tolerance environmental conditions and consideration should be given to having an on-site generator for emergency use.

The relative air pressures between animal areas and service areas of a building housing animals should be considered when the ventilation system is designed to minimize the introduction of airborne disease agents or air pollutants into the service area. Advice of a qualified agricultural engineer or other specialist should be sought for the design of and operating recommendations for ventilation equipment.

AIR QUALITY

Air quality refers to the nature of the air with respect to its effects on the health and well-being of animals and the humans who work with them. Air quality is typically defined in terms of the air content of certain gases, particulates, and liquid aerosols, including those carrying microbes of various sorts. Good ventilation, waste management, and husbandry usually result in acceptable air quality. Ammonia, hydrogen sulfide, carbon monoxide, and methane are the pollutant gases of most concern in animal facilities.

In addition, OSHA has established allowable exposure levels for human workers with 8h of exposure daily to these gases. The concentration of ammonia to which animals are exposed ideally should be less than 10 ppm and should not exceed 25 ppm, but a temporary excess should not adversely affect animal health. Comparable concentrations for hydrogen sulfide are 10 and 50 ppm, respectively. The concentration of carbon monoxide in the air breathed by animals should not exceed 150 ppm, and methane should not exceed 50,000 ppm.

Special ventilation is required when underfloor waste pits are emptied because of the potentially lethal hazards to animals and humans from the hydrogen sulfide and methane gases that are released. Many factors affect airborne dust concentration, including relative humidity, animal activity, air velocity, and type of feed. Dust concentration is lower at higher relative humidities. High animal activity and air velocities stir up more particles and keep them suspended longer.

Fat or oil added to feed reduces dust generation. Microbes and pollutant gases may attach to airborne dust particles. The allowable dust levels specified by OSHA are based on exposure of human workers for 8 h daily without facemasks; allowable dust levels are 5 mg/m3 for respirable dust and 15 mg/m3 for total dust.

Although animals can tolerate higher levels of inert dust with no discernible detriment to their health or well-being, the concentration of dust in animal house air should be minimized. Concentrations of microbes in the air should be minimized. Dust and vapour pressure should be controlled. The ventilation system should preclude the mixing of air from infected microenvironments with that from microenvironments of uninfected animals.

LIGHTING

Lighting should be diffused evenly throughout an animal facility. Illumination should be sufficient to aid in maintaining good husbandry practices and to allow adequate inspection of animals, maintenance of the wellbeing of the animals, and safe working conditions for personnel. Guidelines are available for lighting systems in animal facilities.

Although successful light management schemes are used routinely in various animal industries to support reproductive and productive performance, precise lighting requirements for the maintenance of good health and physiological stability are not known for most animals. However, animals should be provided with both light and dark periods during a 24-h cycle unless the protocol requires otherwise.

Red or dim light may be used if necessary to control vices such as feather-pecking in poultry and tail-biting in livestock. Provision of variable-intensity controls and regular maintenance of light fixtures helps to ensure light intensities that are consistent with energy conservation and the needs of animals, as well as providing adequate illumination for personnel working in animal rooms. A time-controlled lighting system may be desirable or necessary to provide a diurnal lighting cycle. Timers should be checked periodically to ensure their proper operation.

EXCRETA MANAGEMENT AND SANITATION

A complete excreta management system is necessary for any intensive animal facility.

The goals of this system are as follows:

- To maintain acceptable levels of worker health and animal health and production through clean facilities;
- To prevent pollution of water, soil, and air;
- To minimize generation of odours and dust;
- To minimize vermin and parasites;
- To meet sanitary inspection requirements; and
- To comply with local, state, and federal laws, regulations, and policies.

The planning and design of livestock excreta management facilities and equipment are discussed by MWPS. A plan should be followed to ensure that the animals are kept reasonably dry and clean and are provided with comfortable, healthful surroundings. Good sanitation is essential in intensive animal facilities, and principles of good sanitation should be understood by animal care personnel and professional staff.

Different levels of sanitation may be appropriate under different circumstances, depending on whether manure packs, pits, outdoor mounds, dirt floors, or other types of excreta management and housing systems are being used. In some instances, animals may be intentionally exposed to excreta to enhance immunity. A written plan should be developed and implemented for the sanitation of each facility housing agricultural animals.

Building interiors, corridors, storage spaces, anterooms, and other areas should be cleaned regularly and disinfected appropriately. Waste containers should be emptied frequently, and implements should be cleaned frequently. It is good practice to use disposable liners and to wash containers regularly. Animals can harbor microbes that can be pathogenic to humans and other species.

Hence, manure should be removed regularly unless a deep litter system or a builtup manure pack is being employed, and there should be a practical programme of effective disinfection to minimize pathogens in the environment. For terminal cleaning, all organic debris should be removed from equipment and from floor, wall, and ceiling surfaces.

If sanitation depends on heat for effectiveness, the cleaning equipment should be able to supply water that is at least 82°C. When chemical disinfection is used, the temperature of wash water may be cooler. If no machine is available, surfaces and equipment may be washed by hand with appropriate detergents and disinfectants and with vigourous scrubbing.

Health and performance of animals can be affected by the time interval between successive occupations of intensive facilities. Complete disinfection of such quarters during the unoccupied phase of an all-in, all-out regimen of facility management is effective for disease management in some situations. Programmes of pasture-to-crop rotation for periodically resting the pasture and programmes that permit grazing by other animal species can aid in the control of soilborne diseases and parasites. Spreading of manure on pastures as fertilizer is a sound and acceptable management practice but may spread toxic agents and infectious pathogens. Caution should be exercised with manure of animals infected with known pathogens, and other methods of waste disposal should be considered. Animal health programmes should stipulate storage, handling, and use criteria for chemicals designed to inactivate infectious microbes and parasites.

There should be information about prevention, immunization, treatment, and testing procedures for specific infectious diseases endemic in the region. Where serious pathogens have been identified, the immediate environment may need to be disinfected as part of a preventive programme. Elimination of moist and muddy areas in pastures may not be possible, but prolonged destocking is an available option. Drylot facilities may need to be scraped and refilled with uncontaminated materials. Thorough cleaning of animal housing facilities may be followed by disinfection.

Selection of disinfection agents should be based on knowledge of potential pathogens and their susceptibilities to the respective agents. Some means for sterilizing equipment and supplies is essential when certain pathogenic microbes are present and for some specialized facilities and animal colonies. Except in special cases, routine sterilization of equipment, feed, and bedding is not necessary if clean materials from reliable sources are used. In areas where hazardous biological, chemical, or physical agents are being used, a system for monitoring equipment should be implemented.

FEED AND WATER

Animals must be provided with feed and water in a consistent manner, on a regular schedule, in accordance with the requirements established for each species by the NRC and as recommended for the geographic area. When exceptions are required by an experimental or instructional protocol, these must be justified in the protocol and may require approval by the Institutional Animal Care and Use Committee (IACUC). Feeders and waterers must be designed and situated to allow easy access without undue competition.

Sufficient water must be available to meet the animals' daily needs under all environmental conditions. Water troughs, bowls, or other delivery devices must be cleaned as needed to ensure adequate intake and to prevent transmission of microbial-or contaminantassociated disease. Non-municipal water sources should be periodically tested for quality by an approved agency or laboratory. Large supplies of feed should be stored in appropriate, designated areas.

Bulk feed storage containers and feed barrels must be well maintained and the lids kept securely in place to prevent entry of pests, water contamination, and microbial growth. Containers should be cleaned as needed to ensure feed quality. The area around the containers such as the auger boot area should be cleaned regularly. Feed in sacks should be stored off the floor on pallets or racks, and each sack should be labeled with the contents and manufacture date or use-by date.

All feedstuffs should be maintained in such a manner as to prevent contamination by chemicals and/or pests. For example, open feed sacks should be stored in closed containers, and mixing devices and utensils, feed delivery equipment, and feeders/feeding sites should be cleaned regularly to ensure adequate feed intake and prevent transmission of microbial-or contaminant-associated disease. Feed placed in carts or in other delivery devices should be fed promptly or covered to avoid attracting pests.

An effective programme of vermin control should be instituted in feed storage areas. Animal care personnel should routinely inspect feed to identify gross abnormalities such as mold, foreign bodies, or feces; such feed should not be fed until the abnormal components are removed or the feed is determined to be safe. Toxic compounds should be stored in a designated area away from feed and animals to avoid accidental consumption.

SOCIAL ENVIRONMENT

Agricultural animals are social by nature and social isolation is a stressor. Agricultural animals that normally live in herds or flocks under natural conditions that are used in research and teaching should be housed in pairs or groups when possible. Considerations involved in implementing social housing for agricultural animals are discussed by Mench.

If social housing is not feasible because of experimental protocols or because of unpreventable injurious aggression among group members, singly housed animals should be provided with some degree of visual, auditory, and olfactory contact with other members of their species. Socialization to humans and regular positive human contact is also beneficial. In some instances, one species can be used as a companion for another species. Temporary isolation is sometimes required for an animal's safety but the animal should be returned to a social setting as soon as possible.

SEPARATION BY SPECIES

Agricultural animals of different species are typically kept in different enclosures to reduce interspecies conflict, meet the husbandry and environmental needs of the animals, and facilitate research and teaching. However, some research protocols or curricula require species to be co-housed. Facility design and husbandry practices influence whether this can be accomplished in a manner that assures the well-being of the animals.

Mixing of compatible species can often be accomplished more easily in extensive production situations than in intensive housing situations. Some species can carry subclinical or latent infections that can be transmitted to other species that are housed in close proximity, causing clinical disease or mortality. Therefore, a qualified veterinarian or scientist should recommend appropriate health and biosecurity practices if species are to be co-housed.

SEPARATION BY SOURCE OR AGE

Animals obtained from different sources often differ in microbiological status. It is usually desirable to keep these animals separated, at least until microbiologic status is determined or steps are taken to protect against disease transmission. Separation of animals of different ages may also be advisable to reduce disease transmission and control social interactions. Placing animals in groups of similar age or size may allow more uniform access to feed and reduce injuries.

All-in, all-out schemes are examples of age-group separation that are designed to minimize disease risk. However, mixed-group housing is acceptable if disease risk is low, husbandry practices are good, and social interaction is acceptable or necessary. A qualified veterinarian and animal facility manager should work together to devise housing configurations and husbandry practices that assure animal health and well-being while also meeting research and teaching goals.

HUSBANDRY

ANIMAL CARE PERSONNEL

The principal scientist or animal management supervisor should make all animal care personnel aware of their responsibilities during both normal work hours and emergencies. A programme of special husbandry procedures in case of an emergency should be developed. It is the reserach facility management's responsibility to ensure that personnel caring for agricultural animals used for research or teaching are appropriately qualified or trained. This responsibility may be delegated to an IACUC. Qualification by experience and training must be documented.

The animal facility manager must ensure that all animal care personnel are aware of their responsibilities during and outside normal work hours. Protocols for emergency care must be developed and made available to all personnel.

OBSERVATION

Animals in intensive accommodations should be observed and cared for daily by trained and experienced caretakers. Illumination must be adequate to facilitate inspection. In some circumstances, more frequent observation or care may be needed. Under extensive conditions, such as range or pasture, observations should be frequent enough to detect illness or injury in a timely fashion, recognize the need for emergency action, and ensure adequate availability of feed and water. A disaster plan must be developed for observing animals and providing care during emergency weather or health situations.

Regardless of accommodations, animal observations should be documented and husbandry or health concerns reported to the animal facility manager or attending veterinarian as appropriate.

EMERGENCY, WEEKEND, AND HOLIDAY CARE

There must be a means for rapid communication in case of an emergency. In emergencies, facility security and fire personnel must be able to contact staff members responsible for the care of agricultural animals. Names and contact information for those individuals should be posted prominently in the animal facility and provided to the security department or telephone center. If posting names and contact information poses privacy or security issues, a contact number for a security or command center should be used instead.

The institution must ensure that emergency services can be contacted at any time by staff members. The institution must assure continuity of daily animal care, to encompass weekends, holidays, unexpected absences of assigned personnel, and emergency situations. Staff assigned to weekends and holidays must be qualified to perform assigned duties. Cross-training of staff and establishment of standard operating procedures is encouraged to assure consistent, high-quality care.

Emergency veterinary care must be readily available after daily work hours, on weekends, and on holidays. In the event that weather conditions or natural disasters make feeding temporarily impossible, every attempt should be made to provide animals with a continuous supply of water. Absence of feed for up to 48 h should not seriously endanger the health of normal, well-nourished juvenile or adult cattle, sheep, goats, horses, poultry, or swine. Feed should be provided within 24 h to very young animals that are not nursing their dams.

EMERGENCY PLANS

A site-specific emergency plan must be developed to care for agricultural animals that are used for research and teaching. The goal for a plan should be to provide proper management and care for the animals regardless of the conditions. However, some conditions may be so unusual and extreme that it will not be possible to provide immediate care for the animals and to simultaneously ensure employee safety.

Thus, emergency plans should define proper animal management and care and parameters to ensure employee safety. Emergency plans should name employees or positions that are considered essential for providing proper animal management and care. Those employees should a priori understand that responding to emergencies is a condition of employment and that they will be held accountable should they fail to care properly for the animals. Plans should focus on emergencies that are most likely to occur in the specific geographic area or the research or teaching facility. Emergency plans should include animal evacuation plans specific to the research or teaching facility and actions to be taken if transportation is interrupted.

ANIMAL IDENTIFICATION AND RECORDS

Animals should be permanently identified by a method that can be easily read. Identification of individual animals is desirable, but, in some circumstances, it is acceptable to identify animals by group, cage, or pen. Individual birds may be wing-banded or leg-banded. Ear-notching, ear tattooing, electronic transponders, and branding may be used for individual identification of other species, and each has its advantages and disadvantages. Ear notches and tattoos are permanent and effective, but notching constitutes elective surgery and tattoos generally cannot be read without restraining animals.

Electronic transponders require special sensor units or stations, but should be considered when possible. Cattle and horses are most consistently identified using freeze-branding on the hip, shoulder, rear leg, or side. In addition, when freeze branding is used on more than one breed of horse, branding is performed under the mane. Some states require that cattle be permanently identified by branding with a hot iron; however, this procedure is more stressful than freeze-branding.

Ear and neck chain tags, although readable at some distance, can become lost and are therefore not necessarily permanent. In addition, neck chains and straps should be avoided in situations in which the animal could become entangled in a fence, rock outcropping, or other feature of the environment. Any associated pain and distress should be considered when determining the method of identification. In some cases, it may be necessary to identify animals in multiple ways. Individual records are needed for most animals.

These records should include information about the animal, its source and location, its productivity, its reproductive performance, protocols the animal is assigned to, and its ultimate disposition. Records for individual animals or groups should also include dates of vaccination, parasite control measures used, blood testing dates and results, and notations as to whether castration, spaying, or other elective procedures have been performed.

Applicable veterinary data to be recorded include dates of examination/ treatment, clinical information/diagnosis, names of medications and amounts and routes of administration, descriptions of surgical procedures, and resolution of surgical procedures or illnesses. Principal scientists or animal facility managers may wish to record nutritional information. Research protocols often dictate that additional information be recorded.

VERMIN CONTROL

Programmes should be instituted to control infestation of animal facilities by vermin. The most effective control in facilities prevents entry of vermin into the facility by screening openings and ceilings; sealing cracks; eliminating vermin breeding, roosting, and refuge sites; and limiting access of vermin to feed supplies and water sources. Building openings should be screened with 1.3-cm mesh, and ceilings with ridge vents should be screened with 1.9-cm mesh to minimize rodent and bird entry. Smaller mesh sizes are recommended where they will not interfere with airflow.

Mesh may need to be installed along foundations below ground level, especially with wood foundations. Pesticides should be used only as approved. Particular caution should be exercised with respect to residues in feedstuffs, which could injure animals and eventually pass into the meat, milk, or eggs. Pesticides should be used in or around animal facilities only when necessary, only with the approval of the scientist whose animals will be exposed to them, and with special care.

A pesticide applicator or a commercial service may be used. In some regions, wildlife and stray cats and dogs may spread zoonotic diseases, including rabies, to agricultural animals. In highrisk locations, institutions should implement an educational programme that includes training scientific and animal care personnel to recognize the signs of rabies in both wildlife and agricultural species and to handle and report potentially rabid animals.

Inoculation may be advisable for humans who may come into contact with animals in regions where rabies is endemic. Many agricultural institutions keep cats for pest-control purposes. Although the use of free-roaming cats is a traditional form of pest control for agricultural facilities, cats may limit the ability for baiting and may present hygiene or accident risks or serve as disease vectors. However, when cats are present, proper veterinary care and oversight should be provided to these animals. Veterinary care should include vaccinations, parasite control, and neutering.

STANDARD AGRICULTURAL PRACTICES

Sometimes procedures that result in temporary distress and even some pain are necessary to sustain the long-term welfare of animals or their handlers. These practices include comb-, toe-, and beak-trimming of chickens; bill-trimming of ducks; toenail removal, beak-trimming, and snood removal of turkeys; dehorning and hoof-trimming of cattle; taildocking and shearing of sheep; tail-docking, neonatal teeth-clipping, hoof-trimming, and tusk-cutting of swine; and castration of males and spaying of females in some species.

Some of these procedures reduce injuries to humans and other animals. Castration, for example, reduces the chances of aggression against other animals. Bulls and boars also cause many serious injuries to humans. Standard agricultural practices that are likely to cause pain should be reviewed and approved by the IACUC. The development and implementation of alternative procedures less likely to cause pain or distress are encouraged. Overall, best practices for pain prevention and control should be followed.

SICK, INJURED, AND DEAD ANIMALS

Sick and injured animals should be segregated from the main group when feasible, observed thoroughly at least once daily, and provided veterinary care as appropriate. Incurably ill or injured animals in chronic pain or distress should be humanely killed as soon as they are diagnosed as such.

Dead animals are potential sources of infection. Their disposal should be accomplished promptly by a commercial rendering service or other appropriate means and according to applicable ordinances and regulations. Postmortem examination of fresh or wellpreserved animals may provide important animal health information and aid in preventing further losses. When warranted and feasible, waste and bedding that have been removed from facilities occupied by an animal that has died should be moved to an area that is inaccessible to other animals.

SPECIAL CONSIDERATIONS

NOISE

Noise from animals and animal care activities is inherent in the operation of any animal facility. Although differences exist in perceived loudness of the same sound occupational noise limitations have been established for workers, and employees should be provided appropriate hearing protection and monitored for their effects.

Noise ordinarily experienced in agricultural facilities generally appears to have little permanent effect on the performance of agricultural animals, although Algers and Jensen found that continuous fan noise disrupted suckling of pigs. Sudden loud noises have also been reported to cause hysteria in various strains of chickens.

METABOLISM STALLS AND OTHER INTENSIVE PROCEDURES

Animals that are subjected to intensive procedures requiring prolonged restraint, frequent sampling, or other procedures experience less stress if they are trained to cooperate voluntarily with the procedure. Cattle, pigs, and other animals can be trained with food rewards to accept and cooperate with various procedures, such as jugular venipuncture. Many studies of the nutrition and physiology of agricultural animals use a specialized piece of equipment, the metabolism stall. Successful designs have been reported for various species.

These stalls give animal research and care personnel easy access to the animal and its excreta. The degree of restraint of animals housed in metabolism stalls is substantially different from that of other methods that restrict mobility. Animals in metabolism stalls are often held by a head gate or neck tether and are restricted in their lateral and longitudinal mobility. These differences may exacerbate the effects of restriction on animals housed in metabolism stalls. Metabolism stalls should be used only for approved studies, not for the purpose of routine housing.

Researchers should consider appropriate alternatives to metabolism stalls if such alternatives are available. There should be a sufficient preconditioning period to ensure adequate adjustment and comfort of the animal to the metabolism stall before sample collection starts. The length of the preconditioning period should be subject to approval of the IACUC. At least enough space should be provided in the metabolism stall for the animal to rise and lie down normally.

When possible, metabolism stalls should be positioned so that the animal is in visual, auditory, and olfactory contact with conspecific animals to minimize the effects of social isolation. Thermal requirements of animals may be affected when they are placed in metabolism stalls. For example, the lower critical environmental temperature of an animal held individually in a metabolism stall is higher than when residing in a group because the single animal cannot obtain the heat-conserving benefits of huddling with group-mates.

Animals in metabolism stalls should be observed more frequently than those in other environments, and particular attention should be paid to changes in behaviour and appetite and the condition of skin, feet, and legs. The length of time an animal may remain in a metabolism stall before removal for exercise should be based on professional judgment and experience and be subject to approval by the IACUC. The species and the degree of restraint imposed by particular stall types should be taken into consideration in making such judgments.

BIOSECURITY

The term biosecurity in an agricultural setting has historically been defined as the security measures taken to prevent the unintentional transfer of pathogenic organisms and subsequent infection of production animals by humans, vermin, or other means. Biosecurity is also applied in the same context to agricultural animals used in the field of agricultural research, teaching, and testing. With the advent of bioterrorism and the designation of select agents, the term biosecurity has acquired new definitions, depending on the field to which it is applied.

Biosecurity is now used to define national and local policies and procedures that address the protection of food and water supplies from intentional contamination and is additionally used to define measures required to maintain security and accountability of select agents and toxins. It is important to understand these concepts when using the term and to clarify that in this section we are using the term biosecurity in the context of preventing the unintentional transfer of pathogens to animals and humans through appropriate facility design, training, and precautions.

For example, personnel working in swine and poultry facilities should be immunized against influenza and receive training related to potential cross-contamination of agents between animals and humans. The USDA has published voluntary guidelines and a checklist as a resource to help the agricultural producer reduce security risks at the farm level. This publication is designed to prevent both intentional and unintentional introduction of pathogens at the farm level. A list of references and resources is also provided in this document on a variety of farm biosecurity issues.

Other sources of information include reviews of biosecurity basics and good management practices for preventing infectious diseases and biosecurity of feedstuffs. All of these publications offer information and suggestions that could be evaluated for their impact on the design of an animal facility. It is essential that the agricultural animal care staff maintain a high standard of biosecurity to protect the animals from pathogenic organisms that can be transferred by humans. Good biosecurity begins with personal cleanliness.

Showering or washing facilities and supplies should be provided, and personnel should change their clothing as often as necessary to maintain personal hygiene. Disposable gear such as gloves, masks, coats, coveralls, and shoe covers may be required under some circumstances. Personnel should not leave the work place in protective clothing that has been worn while working with animals. Personnel should not be permitted to eat, drink, apply cosmetics, or use tobacco in animal facilities.

Visitors should be limited as appropriate, and institutions should implement appropriate precautions to protect the safety and well-being of the visitors and the animals. Preventing the introduction of disease agents is a continuous challenge, particularly when teaching and research facilities allow public access. Herd health and sanitation programmes should be in place to minimize exposure to pathogens.

Animal care personnel in research and teaching facilities should not be in contact with livestock elsewhere unless strict biosecurity precautions are followed. To reduce inter-building transmission of pathogenic microorganisms, careful attention should be given to traffic patterns of inter-building personnel and disease organisms in feed and transport vehicles.

Barriers to microorganism transmission should be considered for personnel who move between houses, including showering in, changing clothes, and the use of disinfectant footbaths as personnel move between rooms and buildings. Establishing a barrier between animals and visitors requires visitors to do some or all of the following: shower in/shower out, wear clean footwear, change to on-site clothes, and wear only on-site clothes. In addition, if personnel need to go back and forth between different phases of production, it is critical that they work from clean to dirty phases of the farm.

Boot Cleaning and Disinfection

The use of boot baths can prevent or minimize mechanical transmission of pathogens among groups of pigs. Visible organic material may be removed from boots using water and a brush or specific boot cleaning station. Boots may be disinfected by soaking in a clean bath of an appropriate disinfectant following the manufacturer's guidelines for dilution rate and exposure time. Personnel should step into and scrub their boots in the boot bath upon entry and when leaving the room/facility. It is important to frequently empty, clean, and refill the boot bath to prevent it from being contaminated with organic matter. Disposable boots may be used.

BIOCONTAINMENT

High-consequence livestock pathogens or the vectors responsible for transmission of disease cause high morbidity and mortality, and can have a significant regional, national, and global economic impact. The use of these pathogens in agricultural research brings several challenges when designing and operating an animal facility. The design of this type of facility should strive for flexibility, effective containment of pathogens, and minimizing the risk of exposure to personnel when zoonotic agents are utilized.

The use of agricultural animals in high-consequence livestock pathogen research requires a thorough understanding of a variety of regulatory requirements and the concept of risk assessment. The USDA provides a list of livestock, poultry, and fish pathogens that are classified as "pathogens of veterinary significance" *Biosafety in Microbiological and Biomedical Laboratories.* The use of these pathogens requires facilities to meet specific criteria for design, operation, and containment features, which are described in the BMBL.

For the listed agents, criteria may include utilizing containment levels designated as:

- Animal Biosafety Level (ABSL)-2,
- Enhanced ABSL-3,
- BSL-3-Ag, or
- ABSL-4.

Requirements for BSL-3-Ag facilities must be met when any of the listed pathogens are used in animals and the room housing the animals provides the primary containment. When the studies can be accomplished in smaller species in which animals are housed in primary containment devices, which allows the room to serve as the secondary barrier, then enhanced ABSL-3 requirements can be utilized.

Enhancements to ABSL-3 should be determined on a case-by-case basis, using risk assessment, and in consultation with the Animal and Plant Health Inspection Service (APHIS) of the USDA. In addition to the BMBL, facility design standards have been published by the USDA to guide the design of Animal Research Service (ARS) construction projects and contain useful information on the design of containment facilities for agricultural research. These standards include information on containment design that addresses hazard classification and choice of containment, containment equipment, and facility design issues for the different levels of biocontainment. Although published to provide guidance for National Institutes of Health (NIH)-funded construction projects and renovations for biomedical research facilities, the *NIH Design and Policy Guidelines* contain useful information on construction of BSL-3 and ABSL-3 facilities. The use of recombinant DNA molecules in agricultural research can introduce additional considerations when designing an animal facility.

Published guidelines provide recommendations for physical and biological containment for recombinant DNA research involving animals. These guidelines include a supplement published in 2006 that provides additional information specific to the use of lentiviral vectors. The Agricultural Bioterrorism Protection Act of 2002 required the propagation of regulations that address the possession, use, and transfer of select agents and toxins that have the potential to pose a severe threat to plants or animals, and their products.

The USDA/APHIS published the implementing regulation covering animals and animal products, which identifies those select agents and toxins that are a threat solely to animals and animal products and overlap agents, or those agents that pose a threat to public health and safety, to animal health, or to animal products. Overlap select agents and toxins are subject to regulation by both APHIS and the Centers for Disease Control and Prevention.

The regulations implemented by both agencies reference the BMBL and the NIH *Guidelines for Research Involving Recombinant DNA Molecules* as sources to consider when developing physical structure and features, and operational and procedural safeguards. Other issues discussed in some of these references may not directly affect containment of pathogens or safety of personnel, but should be considered as they may affect the design of a facility.

For example, the use of select agents requires certain security measures to be in place that restrict access to areas where select agents or toxins are used or stored. This can include laboratories, animal rooms, and storage freezers, resulting in a significant impact on how a research facility is designed. A thorough understanding of the references cited in this section is advised before initiating the design of new biocontainment facilities or renovation of existing facilities to accommodate research with hazardous agents or toxins requiring containment.

8

Animals Nutrition

Animals need a variety of nutrients to meet their basic needs. These nutrients include fats and carbohydrates that provide energy, proteins that furnish amino acids, vitamins that serve as co-factors for enzymes and perform other functions, ions required for water balance and for nerve and muscle function, and selected elements that are incorporated into certain molecules synthesized by cells.

To determine the levels of nutrients that are needed to sustain normal activities, researchers monitor the relationship between nutrient intake, the levels of nutrients maintained in the body, and health. To determine how nutrition can affect athletic performance, researchers alter the intake of specific nutrients and assess how athletes respond. For example, investigators have shown that athletes who ingest a high-carbohydrate diet for three days outperform individuals who ingest a high-fat, high-protein diet.

There are seven major classes of nutrients: carbohydrates, fats, fibre, minerals, protein, vitamin, and water. These nutrient classes can be categorized as either macronutrients (needed in relatively large amounts) or micronutrients (needed in smaller quantities). The macronutrients are carbohydrates, fats, fibre, proteins, and water. The micronutrients are minerals and vitamins. The macronutrients (excluding fibre and water) provide structural material (amino acids from which proteins are built, and lipids from which cell membranes and some signaling molecules are built) and energy. Some of the structural material can be used to generate energy internally, and in either case it is measured in joules or calories (sometimes called "kilocalories" and on other rare occasions written with a capital C to distinguish them from little 'c' calories).

Carbohydrates and proteins provide 17 kJ approximately (4 kcal) of energy per gram, while fats provide 37 kJ (9 kcal) per gram., though the net energy from either depends on such factors as absorption and digestive effort, which vary substantially from instance to instance. Vitamins, minerals, fibre, and water do not provide energy, but are required for other reasons. A third class dietary material, fibre (*i.e.*, non-digestible material such as cellulose), seems also to be required, for both mechanical and biochemical reasons, though the exact reasons remain unclear. Molecules of carbohydrates and fats consist of carbon, hydrogen, and oxygen atoms. Carbohydrates range from simple monosaccharides (glucose, fructose, galactose) to complex polysaccharides (starch).

Fats are triglycerides, made of assorted fatty acid monomers bound to glycerol backbone. Some fatty acids, but not all, are essential in the diet: they cannot be synthesized in the body. Protein molecules contain nitrogen atoms in addition to carbon, oxygen, and hydrogen. The fundamental components of protein are nitrogen-containing amino acids, some of which are essential in the sense that humans cannot make them internally. Some of the amino acids are convertible (with the expenditure of energy) to glucose and can be used for energy production just as ordinary glucose. By breaking down existing protein, some glucose can be produced internally; the remaining amino acids are discarded, primarily as urea in urine. This occurs normally only during prolonged starvation.

Other micronutrients include antioxidants and phytochemicals which are said to influence (or protect) some body systems. Their necessity is not as well established as in the case of, for instance, vitamins. Most foods contain a mix of some or all of the nutrient classes, together with other substances such as toxins or various sorts. Some nutrients can be stored internally (*e.g.*, the fat soluble vitamins), while others are required more or less continuously. Poor health can be caused by a lack of required nutrients or, in extreme cases, too much of a required nutrient. For example, both salt and water (both absolutely required) will cause illness or even death in too large amounts.

FEEDING BEHAVIOUR

Feeding behaviour, any action of an animal that is directed towards the procurement of nutrients. The variety of means of procuring food reflects the diversity of foods used and the myriad of animal types. The living cell depends on a virtually uninterrupted supply of materials for its metabolism. In multicellular animals the body fluids surrounding each cell are the immediate source of nutrients. The contents of these fluids are kept at a relatively constant level in spite of tolls taken by the cells, primarily by mobilization of nutrients stored in the body; in vertebrates, for example, glucose is stored in the liver, fats in the fat tissues, calcium in the bones. These stores, however, will become exhausted unless the animal takes up nutrients from outside. Movements performed for this purpose are termed feeding behaviour.

NUTRITIONAL REQUIREMENTS OF HIGHER ANIMALS

Cells use nutrients as fuel for energy production (catabolism) and as material for processes of maintenance and growth (anabolism). Multicellular animals derive energy solely from the breakdown of complex organic molecules, mainly carbohydrates and fats. Because the fuel for the maintenance of animal life comes only from other living organisms or their remains, animals are known as heterotrophic organisms. All animal life depends ultimately on the existence of organisms (largely green plants) that can use inorganic sources of energy, of which solar radiation is by far the most important; some microorganisms, however, obtain energy from oxidation of simple inorganic compounds.

For anabolic purposes, food must provide adequate amounts of all chemical elements needed by the cells. Of the approximately 35 elements now known to occur in animal cells, four (oxygen, carbon, hydrogen, and nitrogen) make up about 95 per cent of the cell weight; another nine (calcium, phosphorus, chlorine, sulfur, potassium, sodium, magnesium, iodine, and iron) contribute about 4 per cent. All of these elements have indispensable functions. The remaining 20-odd, together constituting less than 1 per cent of cell weight, are called trace elements, because they occur in minute quantities. Although some of them may become incorporated into cells by accident, many fulfill vital functions.

It is important to note that animal cells cannot synthesize from simple compounds certain necessary complex molecules. Instead, certain large organic molecules must serve as building blocks; such so-called essential dietary components include the vitamins, some amino acids, and certain fatty substances. In general, higher animals appear to have more restricted synthetic powers than lower ones and to require a correspondingly greater number of essential foodstuffs. Microorganisms in the intestines of vertebrates may synthesize materials essential for the host, so that the food of the latter need not contain these substances.

Buffaloes are strict grazers and only browse when feed is utterly scarce. Normally, buffaloes graze during the day. In case of extremely high ambient temperatures, grazing takes place in the morning and afternoon and sometimes during night time. Buffaloes graze more and better than cattle. Thereby they consume more feed and nutrients per kg of body weight than cattle do.

Newborn calves suckle their mothers within two hours of birth. Normal suckling frequency is approximately 6 to 8 times per day. The calves start to nibble grass at 3 to 4 weeks of age, although they are not actually grazing until after a few weeks more. When the calves have reached two months of age, forage starts to become more important than previously and soon most of the nutrient intake comes from forage rather than milk. Natural weaning of calves is usually within a year or before its mother's next parturition.

NUTRITION FOR HORSE

Knowledge of horse nutrition has grown by leaps and bounds during the last 15 years. Research has become more precise and critically evaluated. But more important, this research has given horse owners greater understanding of nutrition. They are more aware of the basic nutrients required by all classes of horses, than in past years. Equine nutrition refers to the feeding of horses, ponies, mules, donkeys, and other equines. Correct and balanced nutrition is a critical component of proper horse care. Horses are herbivores, a type of non-ruminant known as a "hind-gut fermentor." What this means is that horses have only one stomach, similar to humans. However, unlike humans, they also have to digest plant fibre that comes from grass and hay. Therefore, unlike ruminants, who digest fibre in plant matter by use of a multichambered stomach, horses use bacterial fermentation that occurs in the organ known as the *cecum* (or *caecum*) to break down cellulose.

Fig. Grass is a Natural Source of Nutrition for a Horse

In practical terms, horses prefer to eat small amounts of food steadily throughout the day, as they do in nature when grazing on pasture. Although this is not always possible with modern stabling practices and human schedules that favour feeding horses twice a day, it is important to remember the underlying biology of the animal when determining what to feed, how often, and in what quantities. The digestive system of the horse is somewhat delicate.

Because horses are unable to regurgitate food, except from the esophagus, if they overeat or eat something poisonous, vomiting is not an option. They also have a long, complex large intestine and a balance of beneficial bacteria in their cecum that can be upset by rapid changes in feed. Because of these factors, they are very susceptible to colic, which is a leading cause of death in horses. Therefore, horses require clean, high-quality feed, provided at regular intervals, and may become ill if subjected to abrupt changes in their diets. Horses are also sensitive to molds and toxins. For this reason, they must never be fed contaminated fermentable materials such as lawn clippings. Fermented silage, sometimes called "haylage" is fed to horses in some places; however, contamination or failure of the fermentation process that allows any mold or spoilage may be toxic. Horses and other members of the *Equidae* family are adapted by evolutionary biology to eating small amounts of the same kind of food all day long. In the wild, horses ate prairie grasses in semi-arid regions and traveled significant distances each day in order to obtain adequate nutrition. Therefore, their digestive system was made to work best with a small but steady flow of food that does not change much from day to day. Digestion begins in the mouth. First, the animal selects pieces of forage and picks up finer foods, such as grain, with sensitive, prehensile, lips.

The front teeth of the horse, called incisors, clip forage, and food is ground up for swallowing by the premolars and molars. The esophagus carries food to the stomach. The esophagus enters the stomach at an acute angle, creating a one-way valve, with a powerful spincter mechanism at the gastroesophageal junction, which is why horses cannot vomit. The esophagus is also the area of the digestive tract where horses may suffer from choke. Horses have a relatively small stomach for their size, which limits the amount of feed that can be taken in at one time. The average sized horse has a stomach with a capacity of only four gallons, and works best when it contains about two gallons.

One reason continuous foraging or several small feedings per day are better than one or two large meals is because the stomach begins to empty when it is 2/3 full, whether the food in the stomach is processed or not. The small intestine is 50 to 70 feet long and holds 10 to 12 gallons. This is the major digestive organ where most nutrients are absorbed. Bile from the liver acts here, combined with enzymes from the pancreas and small intestine itself. Equids do not have a gall bladder, so bile flows constantly, another reason for a slow but steady supply of food.

Most nutrition is absorbed into the bloodstream from the small intestine. The cecum is the first section of the large intestine. It is also known as the "water gut" or "hind gut." It is a cul-de-sac pouch, about 4 feet long that holds 7 to 8 gallons. It digests cellulose plant fibre through bacterial fermentation. The reason horses must have their diets changed slowly is so the bacteria in the cecum are able to modify and adapt to the different chemical structure of new feedstuffs. Too abrupt a change in diet can cause colic, as the new food is not properly digested.

The large colon, small colon, and rectum make up the remainder of the large intestine. The large colon is 10-12 feet long and holds up to 20 gallons of semi-liquid matter. Its main purpose it to absorb carbohydrates which were broken down from cellulose in the cecum. Due to its many twists and turns, it is a common place for a type of horse colic called an impaction. The small colon is 10-12 feet long, holds about 5 gallons, is the area where the majority of water is absorbed, and where fecal balls are formed. The rectum is about one foot long, and acts as a holding chamber for waste, which is then expelled from the body via the anus.

Nutrients

Like all animals, equines require five main classes of nutrients to survive: water, energy (primarily in the form of fats and carbohydrates), proteins, vitamins, and minerals. Water makes up between 62-68 per cent of a horse's body weight and is essential for life. Horses can only live a few days without water, becoming dangerously dehydrated if they lose 8-10 per cent of their natural body water. Therefore, it is critically important to provide access to a fresh, clean, and adequate supply of water. An average 450kg horse drinks 10 to 12 gallons of water per day, more in hot weather, when eating dry forage such as hay, or when consuming high levels of salt, potassium, and magnesium.

Horses drink less water in cool weather or when on lush pasture, which has a relatively higher water content. When under hard work, or if a mare is lactating, water requirements may be as much as four times greater than normal. Though they need a great deal of water, horses spend very little time drinking; usually 1-8 minutes a day, spread out in 2-8 episodes. Water plays an important part in digestion.

The forages and grains horses eat are mixed with saliva in the mouth to make a moist bolus that can be easily swallowed. Therefore, horses produce up to 10 gallons of saliva per day. Nutritional sources of energy are fat and carbohydrates. Protein is a critical building block for muscles and other tissues. Horses that are heavily exercised, growing, pregnant or lactating need increased energy and protein in their diet. However, if a horse has too much energy in its diet and not enough exercise, it can become too high-spirited and difficult to handle.Fat exists in low levels in plants and can be added to increase the energy density of the diet.

Fat has 9 Mcal/kg of energy, which is 2.25 times that of any carbohydrate source. Because equids have no gall bladder to store large quantities of bile, which flows continuously from the liver directly into the small intestine, fat, though a necessary nutrient, is difficult for them to digest and utilize in large quantities. However, they are able digest a greater amount of fat than can cattle. Horses benefit from up to 8 per cent fat in their diets, but more does not always provide a visible benefit. Horses can only have 15 per cent-20 per cent fat in their diet without the risk of developing diarrhea. Carbohydrates, the main energy source in most rations, are usually fed in the form of hay, grass and grain. Soluble carbohydrates such as starches and sugars are readily broken down to glucose in the small intestine and absorbed. Insoluble carbohydrates, such as fibre (cellulose), are not digested by enzymes, but are fermented by microbes in the cecum and large colon to break down and release their energy sources, the volatile fatty acids.

Soluble carbohydrates are found in nearly every feed source; corn has the highest amount, then barley and oats. Forages normally have only 6 to 8 per cent soluble carbohydrate, but under certain conditions can have up to 30 per cent. Sudden ingestion of large amounts of starch or high sugar feeds can cause colic or laminitis.Protein is used in all parts of the body, especially muscle, blood, hormones, hooves, and hair cells. The main building blocks of protein are amino acids. Alfalfa and other legumes in hay are good sources of protein that can be easily added to the diet. Most adult horses only require 8 to 10 per cent protein in their diet; however, higher protein is important for lactating mares and young growing foals. Horses that are not subjected to hard work or extreme conditions usually have more than adequate amounts of vitamins in their diet if they are receiving fresh, green, leafy forages. Sometimes a vitamin supplement is needed when feeding low-quality hay, if a horse is under stress (illness, traveling, showing, racing, and so on), or not eating well. Grain has a different balance of nutrients than forage, and so requires specialized supplementation to prevent an imbalance of vitamins and minerals.

Minerals are required for maintenance and function of the skeleton, nerves and muscles. These include calcium, phosphorus, sodium, potassium, and chloride, and are commonly found in most good-quality feeds. Horses also need trace minerals such as magnesium, selenium, copper, zinc and iodine. Normally, if adult animals at maintenance levels are consuming fresh hay or are on pasture, they will receive adequate amounts of minerals in their diet, with the exception of sodium chloride (salt), which needs to be provided, preferably free choice.

Some pastures are deficient in certain minerals, especially selenium or copper, and in such situations, deficiency diseases may occur if the horses' feed is not supplemented. Calcium and phosphorus are needed in a specific ratio of between 1:1 and 2:1. Adult horses can tolerate up to a 5:1 ratio, foals no more than 3:1. A total ration with a higher ratio of phosphorus than calcium is to be avoided. Over time, the imbalance will ultimately lead to bone problems such as osteoporosis. Foals and young growing horses through their first three to four years have special nutritional needs and require feeds that are balanced with a proper calcium:phosphorus ratio and other trace minerals. A number of skeletal problems may occur in young animals with an unbalanced diet. Hard work increases the need for minerals; sweating depletes sodium, potassium, and chloride from the horse's system. Therefore, supplementation with electrolytes may be required for horses in intense training, especially in hot weather.

TYPES OF FEED

Equids can consume approximately 2 per cent to 2.5 per cent of their body weight in dry feed each day. Therefore, a 1000 lb adult horse could eat up to 25 pounds of food. Foals less than six months of age eat 2 to 4 per cent of their weight each day.Solid feeds are placed into three categories: forages, concentrates (including grain or pelleted rations), and supplements. Equine nutritionists recommend that 50 per cent or more of the animal's diet by weight should be forages. If a horse is working hard and requires more energy, the use of grain is increased and the percentage of forage decreased so that the horse obtains the energy content it needs for the work it is performing. However, forage amount should never go below 1 per cent of the horse's body weight per day.

Forages, also known as "roughage," are plant materials classified as legumes or grasses, found in pastures or in hay. Often, pastures and hayfields will contain a bland of both grasses and legumes. Nutrients available in forage vary greatly with maturity of the grasses, fertilization, management, and environmental conditions. Grasses are tolerant of a wide range of conditions and contain most necessary nutrients. Some commonly used grasses include timothy, brome, fescue, coastal bermuda, orchard grass, and Kentucky bluegrass. Legumes such as clover or alfalfa are usually higher in protein, calcium, and energy than grasses.

However, they require warm weather and good soil to produce the best nutrients. Legume hays are generally higher in protein than the grass hays. They are also higher in minerals, particularly calcium, but have an incorrect ratio of calcium to phosphorus. Because they are high in protein, they are very desirable for growing horses or those subjected to very hard work, but the calcium:phosphorus ratio must be balanced by other feeds to prevent bone abnormalities. Hay is a dried mixture of grasses and legumes. It cut in the field and then dried and baled for storage. Hay is most nutritious when it is cut early on, before the seed heads are fully mature and before the stems of the plants become tough and thick.

Hay that is very green can be a good indicator of the amount of nutrients in the hay; however, colour should not be used as sole indicator - smell and texture are also important. Hay can be analyzed by many laboratories and that is the most reliable way to tell the nutritional values it contains. Moldy or dusty hay should not be fed to horses, as it is the most common cause of Recurrent airway obstruction, also known as COPD or "heaves." Hay, particularly alfalfa, is sometimes compressed into pellets or cubes. Processed hay can be of more consistent quality and is more convenient to ship and to store. It is also easily obtained in areas that may be suffering localized hay shortages.

However, these more concentrated forms can be overfed and horses are somewhat more prone to choke on them. On the other hand, hay pellets and cubes can be soaked until they break apart into a pulp or thick slurry, and in this state are a very useful source of food for horses with tooth problems such as dental disease, tooth loss due to age, or structural anomalies.

Haylage, also known as *Round bale silage* is a term for grass sealed in airtight plastic bags, a form of forage that is frequently fed in the United Kingdom and continental Europe, but is not often seen in the United States. Because haylage is a type of silage, hay stored in this fashion must remain completely sealed in plastic, as any holes or tears can stop the preservation properties of fermentation and lead to mold or spoilage. Rodents chewing through the plastic can also spoil the hay introducing contamination to the bale.If a rodent dies inside the plastic, the subsequent botulism toxins released can contaminate the entire bale. Another type of forage sometimes provided to horses is beet pulp, a byproduct left over from the processing of sugar beets, which is high in energy as well as fibre.Sometimes, straw or chaff from cereal grains is fed to animals. However, this is roughage with little nutritional value. It is not recommended by nutritionists as a horse feed, though it is sometimes used as a filler; it can slow down horses who eat their grain too fast, or it can provide additional fibre when the horse must meet most nutritional needs via concentrated feeds. Straw is more often used as a bedding in stalls to absorb wastes.

Whole or crushed grains are the most common form of concentrated feed, sometimes referred to generically as "oats" or "corn" even if those grains are not present, also sometimes called "straights" in the UK. Oats are the most popular grain for horses. Oats have a lower digestible energy value and higher fibre content than most other grains. They form a loose mass in the stomach that is well suited to the equine digestive system. They are also more palatable and digestible other grains. Corn, referred to as Maize in the UK, is the second most palatable grain. It provides twice as much digestible energy as an equal volume of oats and is low in fibre. Because of these characteristics, is easy to over-feed corn, causing obesity, so horses are seldom fed corn all by itself. Nutritionists caution horse owners that moldy corn should never be fed because it is poisonous to horses. Barley is also fed to horses, but needs to be processed to crack the seed hull and allow easier digestibility. It is frequently fed in combination with oats and corn, a mix informally referred to by the acronym "COB" (for Corn, Oats and Barley).

Wheat is generally not used as a concentrate. However, wheat bran is sometimes added to the diet of a horse for supplemental nutrition, usually moistened and in the form of a bran mash. Wheat bran is high in phosphorus, so must be fed carefully so that it does not cause an imbalance in the Ca:P ratio of a ration. Once touted for a laxative effect, this use of bran is now considered unnecessary, as horses, unlike humans, obtain sufficient fibre in their diets from other sources. Many feed manufacturers combine various grains and add additional vitamin and mineral supplements to create a complete premixed feed that is easy for owners to feed and of predictable nutritional quality. Some of these prepared feeds are manufactured in pelleted form, others retain the grains in their original form.

In many cases molasses as a binder to keep down dust and for increased palatability. Grain mixes with added molasses are usually called "Sweet feed" in the USA and "Coarse mix" in the United Kingdom. Pelleted or *extruded* feeds (sometimes referred to as "nuts" in the UK) may be easier to chew and result in less wasted feed. Horses generally eat pellets as easily as grain. However, pellets are also more expensive, and even "complete" rations do not eliminate the necessity for forage. The average modern horse on good hay or pasture with light work usually does not need supplements; however, horses subjected to stress due to age, intensive athletic work, or reproduction may need additional nutrition. Extra fat and protein are sometimes added to the horse's diet, along with vitamin and mineral supplements.

Soybean meal is a common protein supplement, and averages about 44 per cent crude protein. The protein in soybean meal is high-quality, with the proper ratio of dietary essential amino acids for equids. Cottonseed meal, Linseed meal, and peanut meal are also used, but are not as common. Vegetable oil is a common fat source added to a ration. Corn oil is particularly popular, but other oils are used as well. Rice bran is a very good fat supplement that contains 20 per cent fat as well as fibre and other nutrients. Flax seed is another good source of fat, though it must ground up for horses to digest it. Some commercial feed manufacturers now make products containing both flaxseed and rice bran.

There are hundreds, if not thousands of commercially prepared vitamin and mineral supplements on the market, many tailored to horses with specialized needs. Most horses only need quality forage, water and a salt or mineral block. Grain or other concentrates are often not necessary. But, when grain or other concentrates are fed, quantities must be carefully monitored. To do so, horse feed is measured by weight, not volume. For example, 1 lb. of oats has a different volume than 1 lb. of corn. When continuous access to feed is not possible, it is more consistent with natural feeding behaviour to provide three small feedings per day instead of one or two large ones.

However, even two daily feedings is preferable to only one. To gauge the amount to feed, a weight tape, available at most feed stores, can be used to provide a reasonably accurate estimate of a horse's weight. The tape measures the circumference of the horse's barrel, just behind the withers and elbows, and the tape is calibrated to convert inches or centimeters into approximate pounds or kilograms.

Actual amounts fed vary by the size of the horse, the age of the horse, the climate, and the work to which the animal is put. In addition, genetic factors play a role. Some animals are naturally easy keepers, which means that they can thrive on relatively small amounts of food and are prone to obesity and other health problems if overfed. Others are hard keepers, meaning that they are prone to be thin and require considerably more food to maintain a healthy weight.

Veterinarians are usually a good source for recommendations on appropriate types and amounts of feed for a specific horse. There are also numerous books written on the topic. Feed manufacturers usually offer very specific guidelines for how to select and properly feed products from their company, and in the United States, the local office of the Cooperative Extension Service can provide educational materials and expert recommendations. Equids always require forage. When possible, nutritionists recommend it be available at all times, at least when doing so does not overfeed the animal and lead to obesity. It is safe to feed a ration that is 100 per cent forage (along with water and supplemental salt), and any feed ration should be at least 50 per cent forage. Hay with alfalfa or other legumes has more concentrated nutrition and so is fed in smaller amounts than grass hay, though many hays have a mixture of both types of plant.

When beet pulp is fed, a ration of 2 to 5 pounds is usually soaked in water for 3 to 4 hours prior to feeding in order to make it more palatable, and to minimize the risk of choke and other problems. It is usually soaked in a proportion of one part beet pulp to two parts water. Beet pulp is usually fed in addition to hay, but occasionally is a replacement for hay when fed to very old horses who can no longer chew properly. Some pelleted rations are designed to be a "complete" feed that contains both hay and grain, meeting all the horse's nutritional needs. However, even these rations should have some hay or pasture provided, a minimum of a half-pound of forage for every 100 pounds of horse, in order to keep the digestive system functioning properly and to meet the horse's urge to graze.

Recent studies address the level of various non-structural carbohydrates (NSC), such as fructan, in forages. Too high an NSC level causes difficulties for animals prone to laminitis or equine polysaccharide storage myopathy (EPSM). NSC cannot be determined by looking at forage, but hay and pasture grasses can be tested for NSC levels. Concentrates, when fed, are recommended to be provided in quantities no greater than 1 per cent of a horse's body weight per day, and preferably in two or more feedings. If a ration needs to contain a higher per cent of concentrates, such as that of a race horse, bulky grains such as oats should be used as much as possible; a loose mass of feed helps prevent impaction colic. Peptic ulcers are linked to a too-high concentration of grain in the diet, particularly noticed modern racehorses, where some studies show such ulcers affecting up to 90 per cent of all race horses.

In general, the portion of the ration that should be grain or other concentrated feed is 0 to 10 per cent grain for mature idle horses; between 20 to 70 per cent for horses at work, depending on age, intensity of activity and energy requirements. Concentrates should not be fed to horses within one hour before or after a heavy workout. Concentrates also need to be adjusted to level of performance. Not only can excess grain and inadequate exercise lead to behaviour problems, it may also trigger Equine Exertional Rhabdomyolysis, or "tying up," in horses prone to the condition. Horses normally require free access to all the fresh, clean water they want, and to avoid dehydration, should not be kept from water longer than four hours at any one time. However, water may need to be temporarily limited in quantity when a horse is very hot after a heavy workout. As long as a hot horse continues to work, it can drink its fill at periodic intervals, provided that common sense is used and that an overheated horse is not forced to drink from extremely cold water sources.

But when the workout is over, a horse needs to be cooled out and walked for 30—90 minutes before it can be allowed all the water it wants at one time. However, dehydration is also a concern, so some water needs to be offered during the cooling off process. A hot horse will properly rehydrate while cooling off if offered a few swallows of water every three to five minutes while being walked. Sometimes the thirst mechanism does not immediately kick in following a heavy workout, which is another reason to offer periodic refills of water throughout the cooling down period.

Even a slightly dehydrated horse is at higher risk of developing impaction colic. Additionally, dehydration can lead to weight loss because the horse cannot produce adequate amounts of saliva, thus decreasing the amount of feed and dry forage consumed. Thus, it is especially important for horse owners to encourage their horses to drink when there is a risk of dehydration; when horses are losing a great deal of water in hot weather due to strenuous work, or in cold weather due to horses' natural tendency to drink less when in a cold environment. To encourage drinking, owners may add electrolytes to the feed, additives to make the water especially palatable (such as apple juice), or, when it is cold, to warm the water so that it is not at a near-freezing temperature.

SPECIAL FEEDING ISSUES FOR PONIES

Ask a pony owner and they'll tell you, ponies are tougher, smarter, stronger and healthier than any horse. They run on a lot less fuel too. Rations that would starve a horse will keep a pony round and energetic. Most pony breeds developed where the pasture was sparse, the terrain rugged, and the climate harsh. When we pamper them, feed special preparations and lush grass we sometimes do more harm than good. Ponies need only the fraction of the feed that horses do.

Hay for ponies should be good quality grass hay. Your pony probably won't need the nutrition provided by alfalfa and clovers. Lush pasture is a danger zone for ponies. Ponies can founder in less than 60 minutes of grazing if introduced suddenly to lush grass. If you plan to keep your pony on grass introduce it very slowly. Start with 10 minutes of grazing and gradually add a few minutes each day twice a day. If your pony eats too much rich pasture it could lead to such things as colic or founder. You may never be able to leave some ponies on good pasture. Ponies can become obese very quickly and that can lead to health problems.

Good pasture for a pony would be one that he has to work at finding the grass in. Sparse grass that grows slowly would be ideal. Or he could spend a small portion of his time on pasture and the rest in a grassless paddock. Some people use their round pen or a paddock where no grass grows. Another option is to use a grazing muzzle. Make sure your pony has access to clean fresh water. Ponies rarely need concentrates or grain. The exception would be a pony that is working hard: one that is doing several lessons a week, is being driven frequently, is doing something like pulling competitions, or a lactating mare. If your pony is losing condition you could increase the quantity of hay and if that isn't enough add a concentrate that isn't too rich. A forage replacer fortified with vitamins and minerals might give your pony the nutritional boost it needs. Pony mouths are small, so overgrown teeth can be a problem. Don't forget to have your vet check your pony's teeth to ensure it can chew easily. Ponies also need regular de-worming to keep them in the best health.

If you like to feed your ponies even if it is not working hard--and for some owners this is a very satisfying activity--look for a concentrate that is low in calories. Some manufacturers make special pony mixes. These mixes are balanced with the correct amount of supplements for a pony. Don't be tempted to top dress it with a lot of extras like molasses or beet pulp. If you are feeding good hay, the pony is getting a bit of pasture and you have a mineral/salt block available your pony will be getting what he needs. Ponies and miniature horses are usually easy keepers and need less feed than full-sized horses.

This is not only because they are smaller, but also, because they evolved under harsher living conditions than horses, they use feed more efficiently. Ponies easily become obese from overfeeding and therefore are at high risk for colic, Equine Metabolic Syndrome, and, especially, laminitis. Fresh grass is a particular danger to ponies; they can develop laminitis in as little as one hour of grazing on lush pasture.

It is important to track the weight of a pony carefully, by use of a weight tape. Forages may be fed based on weight, at a rate of about 1 pound of forage for every 100 pounds. Forage, along with water and a salt and mineral block, is all most ponies require. If a hard-working pony needs concentrates, a ratio of no more than 30 per cent concentrates to 70 per cent forage is recommended. Concentrates designed for horses, with added vitamins and minerals, will often provide insufficient nutrients at the small serving sizes needed for ponies. Therefore, if a pony requires concentrates, feed and supplements designed specially for ponies should be used. In the UK, extruded pellets designed for ponies are sometimes called "pony nuts.".

Special Feeding Issues for Mules and Donkeys

Like ponies, mules and donkeys are also very hardy and generally need less concentrated feed than horses. Mules need less protein than horses and do best on grass hay with a vitamin and mineral supplement. If mules are fed concentrates, they only need about half of what a horse requires. Like horses, mules require fresh, clean water, but are less likely to over-drink when hot.

Fig. Grass hay for Mules and Donkey

Donkeys, like mules, need less protein and more fibre than horses. They do best when allowed to consume small amounts of food over long periods, as is natural for them in an arid climate. They can meet their nutritional needs on 6 to 7 hours of grazing per day on average dryland pasture that is not stressed by drought. If they are worked long hours or do not have access to pasture, they require hay or a similar dried forage, with no more than a 1:4 ratio of legumes to grass. They also require salt and mineral supplements, and access to clean, fresh water. Like ponies and mules, in a lush climate, donkeys are prone to obesity and are at risk of laminitis. Many people like to feed horses special treats such as carrots, sugar cubes, peppermint candies or specially manufactured horse "cookies." Horses do not need treats, and due to the risk of colic or choke, many horse owners do not allow their horses to be given treats. There are also behavioural issues that some horses may develop if given too many treats, particularly a tendency to bite if hand-fed, and for this reason many horse trainers and riding instructors discourage the practice.

However, if treats are allowed, carrots and compressed hay pellets are common, nutritious, and generally not harmful. Apples are also acceptable, though it is best if they are first cut into slices. Horse "cookies" are often specially manufactured out of ordinary grains and some added molasses. They generally will not cause nutritional problems when fed in small quantities. However, many types of human foods are potentially dangerous to a horse and should not be fed. This includes bread products, meat products, candy, and carbonated or alcoholic beverages.

It was once a common practice to give horses a weekly *bran mash* of wheat bran mixed with warm water and other ingredients. It is still done regularly in some places. While a warm, soft meal is a treat many horses enjoy, and was once considered helpful for its laxative effect, it is not nutritionally necessary. An old horse with poor teeth may benefit from food softened in water, a mash may help provide extra hydration, and a warm meal may be comforting in cold weather, but horses have far more fibre in their regular diet than do humans, and so any assistance from bran is unnecessary. There is also a risk that too much wheat bran may provide excessive phosphorus, unbalancing the diet, and a feed of unusual contents fed only once a week could trigger a bout of colic. All hay and concentrated feeds must kept dry and free of mold, rodent feces and other types of contamination that may cause illness in horses. Feed kept outside or otherwise exposed to moisture can develop mold quite quickly. Due to fire hazards, hay is often stored under an open shed or under a tarp, rather than inside a horse barn itself, but should be kept under some kind of cover.

Concentrates take up less storage space, are less of a fire hazard, and are usually kept in a barn or enclosed shed. A secure door or latched gate between the animals and any feed storage area is critical. Horses accidentally getting into stored feed and eating too much at one time is a common but preventable way that horses develop colic or laminitis. It is also important to never give a horse feed that was contaminated by the remains of a dead animal, it is a potential source of botulism. This is not an uncommon situation. For example, mice and birds can get into poorly stored grain and be trapped; hay bales sometimes accidentally contain snakes, mice, or other small animals that were caught in the baling machinery during the harvesting process.

Horses can become anxious or stressed if there are long periods of time between meals. They also do best when they are fed on a regular schedule, they are creatures of habit and easily upset by changes in routine. When horses are in a herd, their behaviour is hierarchical; the higher-ranked animals in the herd eat and drink first. Low-status animals, who eat last, may not get enough food, and if there is little available feed, higher-ranking horses may keep lower-ranking ones from eating at all. Therefore, unless a herd is on pasture that meets the nutritional needs of all individuals, it is important to either feed horses separately, or spread feed out in separate areas to be sure all animals get roughly equal amounts of food to eat. In some situations where horses are kept together, they may still be placed into separate herds, depending on nutritional needs; overweight horses are kept separate from thin horses so that rations may be adjusted accordingly.

Horses' teeth continually erupt throughout their life, are worn down as they eat, and can develop uneven wear patterns that can interfere with chewing. For this reason, horses need a dental examination at least once a year, and particular care must be paid to the dental needs of older horses. The process of grinding off uneven wear patterns on a horse's teeth is called *floating* and can be performed by a veterinarian or a specialist in equine dentistry.Diet and Nutrition for Dog There are a great quantity of commercial foods and treats marketed for dogs, and not all are recommended as part of a balanced, healthy diet.

There is some debate as to whether domestic dogs should be classified as omnivores or carnivores, by diet. The classification in the Order Carnivora does not necessarily mean that a dog's diet must be restricted to meat; unlike an obligate carnivore, such as the cat family with its shorter small intestine, a dog is neither dependent on meat-specific protein nor a very high level of protein in order to fulfill its basic dietary requirements. Dogs are able to healthily digest a variety of foods including vegetables and grains, and in fact dogs can consume a large proportion of these in their diet. Wild canines not only eat available plants to obtain essential amino acids, but also obtain nutrients from vegetable matter from the stomach and intestinal contents of their herbivorous prey, which they usually consume.

Domestic dogs can survive healthily on a reasonable and carefully designed vegetarian diet, particularly if eggs and milk products are included. Some sources suggest that a dog fed on a strict vegetarian diet without L-carnitine may develop dilated cardiomyopathy, however, L-carnitine is found in many nuts, seeds, beans, vegetables, fruits and whole grains. In the wild, dogs can survive on a vegetarian diet when animal prey is not available.

Observation of extremely stressful conditions such as the Iditarod Trail Sled Dog Race, and scientific studies of similar conditions has shown that high-protein (approximately 40 per cent) diets including meat help prevent damage to muscle tissue in dogs and some other mammals. This level of protein corresponds to the percentage of protein found in the wild dog's diet when prey is abundant; higher levels of protein seem to confer no added benefit.

Dogs frequently eat grass, which is a harmless activity. Explanations abound, but rationales such as that it neutralizes acid, or that dogs eat grass to induce vomiting to remove unwanted substances from their stomachs, are at best educated guesses. Dogs do vomit more readily than humans, as part of their typical feeding behaviour of gulping down food then regurgitating indigestible material such as bones and fur. This behaviour is typical of pack feeding in the wild, where the most important thing is to get as much of the kill as possible before others consume it all. Individual domestic dogs, however, may be very "picky" eaters, in the absence of this social pressure. Dogs may also appear to eat grass when they are just running the blades through their mouth to gather information. Their sense of smell and taste may act together to detect if other animals have walked through their area or urinated on the grass.

DANGEROUS SUBSTANCES

Human Food

Some foods commonly enjoyed by humans are dangerous to dogs, including chocolate (Theobromine poisoning), onions, grapes and raisins, some types of gum, certain sweeteners and Macadamia nuts. The only known dangerous substance in chocolate is cocoa, so the danger of white chocolate is uncertain.

The acute danger from grapes and raisins was discovered around 2000, and has slowly been publicized since then. The cause is not known. Small quantities will induce acute renal failure. Sultanas and currants may also be dangerous. Cooked bones are dangerous for dogs, because the heat of cooking changes their chemical and physical properties so that they cannot be chewed properly. As a result they may splinter into jagged shards that resist digestion. Alcoholic beverages pose comparable hazards to dogs as they do to humans, but due to low body weight and lack of alcohol tolerance they are toxic in much smaller portions.

Household Poisons

Many household cleaners such as ammonia, bleach, disinfectants, drain cleaner, soaps, detergents, and other cleaners, mothballs and matches are dangerous to dogs, as are cosmetics such as deodorants, hair colouring, nail polish and remover, home permanent lotion, and suntan lotion. Dogs find some poisons attractive, such as antifreeze (automotive coolant), slug and snail bait, insect bait, and rodent poisons. Antifreeze is insidious to dogs, either puddled or even partly cleaned residue, because of its sweet taste. A dog may pick up antifreeze on its fur and then lick it off.

Plants

Plants such as caladium, dieffenbachia and philodendron will cause throat irritations that will burn the throat going down as well as coming up. Hops are particularly dangerous and even small quantities can lead to malignant hyperthermia.

Amaryllis, daffodil, english ivy, iris, and tulip (especially the bulbs) cause gastric irritation and sometimes central nervous system excitement followed by coma, and, in severe cases, even death. Ingesting foxglove, lily of the valley, larkspur and oleander can be life threatening because the cardiovascular system is affected. Yew is very dangerous because it affects the nervous system. Immediate veterinary treatment is required for dogs that ingest these.

Animal Feces

Dogs occasionally eat their own feces, or the feces of other dogs and other species if available, such as cats, deer, cows, or horses. This is known as coprophagia.

Some dogs develop preferences for one type over another. There is no definitive reason known, although boredom, hunger, and nutritional needs have been suggested. Eating cat feces is common, possibly because of the high protein content of cat food. Dogs eating cat feces from a litter box may lead to to Toxoplasmosis. Dogs seem to have different preferences in relation to eating feces. Some are attracted to the stools of deer, cows, or horses.

Other risks

Human medications may be toxic to dogs, for example paracetamol/ acetaminophen (Tylenol). Zinc toxicity, mostly in the form of the ingestion of US pennies minted after 1982, is commonly fatal in dogs where it causes a severe hemolytic anemia.

COW DIET

Now what, necessarily, do cows eat? Calves, when first born, are fed milk, much like human infants. Unlike human babies, however, calves can walk within an hour of being born, and three days later they are introduced to calf grain. Whether or not they start eating it is up to them. Calf grain is mainly comprised of corn followed by other grain products, such as wheat and oats. And, to sweeten the deal, molasses is added, giving the grain a sweet smell. As humans grow, so do their appetites, which are the same as those of dairy cows.

During the four- to five-week range, calves are introduced to hay and cow feed (as opposed to calf feed). Hay is dried alfalfa — so in human terms, it's a salad! Then in the six-to eight-week range, dairy farmers will wean calves off milk — that is stop feeding it to them and start pushing more solids and water.

Now the calves become young adults or teenagers, and they're called heifers. Heifers are put together in a group until they become pregnant. Once a heifer has her first calf, she becomes a cow. In order for a cow to continue to produce milk, she has to have a baby every year. Adult cows are like adult humans — they have a job.

Theirs is to produce milk. Cows are fed a number of things, including cow feed, hay, grass (in the spring and summer) and silage. Silage can be either corn or alfalfa. Corn silage is slightly different because it contains the whole stalk of corn. Of course, cows receive tons and tons of water, as it takes five gallons of water to produce one gallon of milk. Happy cows equal more milk! Corn has been a staple ingredient in lactating dairy cow rations for a hundred years. However, many dairy farms import corn to feed to their dairy cows.

The price of this imported corn has been fairly stable for a number of years. The government is trying to reduce dependence on foreign oil by supporting efforts to produce fuel (ethanol) from renewable resources like corn or cellulose. There is widespread public support for incorporating ethanol into fuelburning vehicles (gasoline and diesel) because ethanol addition results in greater combustion of these fuels and is nontoxic and biodegradable in the water and the soil.

Fig. Feeding Dairy Cow

The Congress passed legislation called a renewable fuels standard (RFS) that will at least double the use of ethanol and biodiesel by the year 2012. The number of plants producing ethanol from corn is projected to increase dramatically. This growth in ethanol production has reduced the amount of corn available for livestock feed and concurrently increased its market price. Increased corn prices have resulted in increased feed costs. While higher milk prices may compensate for ration cost increases, partially replacing corn with other less expensive feedstuffs may be required over the longer term.

ALTERNATE RATION STRATEGIES

Other Grain Sources of Dietary Starch

Fig. Grain Diet for cow

Grains, like corn, are an excellent source of starch which is highly fermentable by ruminal microorganisms. The propionic acid produced by starch-fermenting bacteria is converted to glucose by the cow's liver; this glucose is used to make milk. In addition, the bacteria provide about 50 to 60 per cent of the protein needs of the cow as they are washed out of the rumen and are digested in the abomasum and small intestine. To optimize milk production, starch should make up 24 to 26 per cent of the dietary dry matter.

Barley and wheat are rich sources of dietary starch. Shifting cow diets between corn and barley or wheat needs to be done slowly.

The starch in barley and wheat is more rapidly available than that in corn so quick dietary changes can result in digestive upsets and lowered ruminal pH. Wheat flour can sometimes be bought for less than corn, but it is very dusty and potentially explosive so it is often avoided by mills. Field reports suggest that flour might 'paste up" in the rumen, so it is not a popular feed ingredient. Hominy contains less starch but more protein, fibre, and fat compared to corn, however their energy density is similar. Replacing all of the corn with hominy will reduce the dietary starch concentration about 3 percentage units. Dietary costs would be expected to be reduced by replacing corn with hominy. If no milk is lost, then profit is improved. Experiments comparing performance of cows fed ground corn versus hominy could not be located. However, many Florida dairies feed hominy successfully. Even replacing half the corn with hominy can be a reasonable strategy. Because hominy is higher in fat and phosphorus than corn, care should be taken to avoid overfeeding these nutrients.

Low Starch Feedstuffs as Possible Substitutes for Corn

Because grain sources rich in starch, such as barley or wheat, may not be economically available year-round, other commodity feeds must be considered. Ingredients with lower prices than corn should be considered as partial replacements for corn. Most common feedstuffs contain quite a bit less starch than corn. The best starch sources after corn include corn silage at 30 per cent, wheat midds at 26 per cent, wheat bran at 23 per cent, rice bran at 19 per cent, and corn gluten feed at 16 per cent (DM basis). All other feedstuffs are less than 10 per cent starch. Replacing some of the corn with these feeds will reduce dietary starch to less than 25 per cent which is a common, US ration target. How far can dietary starch be reduced without significantly affecting milk production?

Corn Feed

Corn gluten feed (CGF) is a byproduct of the manufacture of corn sweeteners, corn starch, corn syrup, and corn oil using the wet milling process. The corn starch is used to make ethanol. The CGF consists of the corn bran and a steep liquor (fermented nutrients extracted from water used to soak corn grain) mixed in approximately a 2:1 ratio.

CGF is sold in both wet and dry forms. It contains 16 per cent starch, 36.1 per cent NDF, and 23.5 per cent protein, the protein being highly degradable in the rumen. Pricing is substantially lower than corn. CGF may serve as a concentrate, replacing only the corn and soybean meal, or as both a concentrate and a fibre source, replacing both concentrate and traditional forages. CGF generally accounted for 10 to 45 per cent of the diet, although some studies went higher. When CGF was fed at about 20 per cent of ration DM, milk production was decreased statistically in 1 study, increased in 1 study, and remained unchanged in 9 other studies. Cows in these studies were milking between 50 and 90 pounds per day.

Increasing the CGF to 30, 40 or even 57 per cent of the diet did not reduce milk in any study with the exception of Staples *et al*. In two studies, in which CGF was fed between 38 and 40 per cent of the diet replacing both forage and concentrate starting at calving, average milk yield was increased significantly by feeding the corn gluten feed. As more CGF replaced more corn, the dietary starch concentrations dropped, going as low as 15 per cent in some cases. In spite of lowered starch intake, milk yield did not drop significantly.

The bulk of these studies indicate that dietary starch could be reduced by replacing some corn and protein meals with the digestible fibre found in CGF. Dropping dietary starch from 26 to 21 per cent may be an acceptable compromise. Sample diets were formulated to include CGF at 0 per cent, 10 per cent, and 20 per cent of the diet; starch was reduced concurrently from 25.9 to 23.5 per cent to 20.9 per cent in these rations. As CGF increases in diet, the proportion of extruded soy meal increased to keep the ruminally undegradable protein constant. Degradable protein sources had to be reduced to keep the dietary crude protein from exceeding 17 per cent. Whether this approach will maintain milk production is not known. Milk and feed intake will be maintained when CGF is fed at 20 per cent of dietary DM. But feeding CGF at 10 per cent of the diet may be a preferred approach initially to "play it safe." Some signs that may indicate that the dietary starch has gotten too low include; dropped milk production, stiffer manure, increases in milk urea nitrogen, and loss of body condition.

Another concern is that drying CGF can reduce the digestibility of the protein if the drying temperature is too high. This is determined by analyzing CGF for acid detergent insoluble protein (ADIN). Bernard *et al*.reported that dried CGF had higher concentrations of ADIN than wet CGF. In recent years, high ADIN may not be as big of a problem in CGF based upon analyses reported by Dairy One. The ADIN values were less than 5 per cent of the nitrogen in the majority of samples; ADIN is usually not considered a problem until it reaches 10 per cent of the total nitrogen in the feed. Commodities need to have some consistent level of nutrient density from delivery to delivery in order to keep the daily ration nutrients consistent for the cows. Dairy One lists on their web site the normal range and standard deviation of the nutrients of most of the feeds they have analysed over the past 6 years.

Corn has an average starch value of 70.6 per cent. The standard deviation is 5.1 per cent. This means that two-thirds of the samples that were analysed ranged between 65.5 per cent and 75.7 per cent which is 5.1 per cent units added to or subtracted from the mean of 70.6 per cent. This also means that one-third of the samples analysed were outside this range. This is a fairly narrow range compared to byproduct feeds. Starch in CGF samples has a typical range of 8.6 per cent to 24.0 per cent with a mean of 16.3 per cent. This large variation is likely due to the fact that CGF production is not standardized across mills. Likewise, starch in hominy typically ranges between 42.9 per cent and 63.9 per cent with a mean of 53.4 per cent. By feeding these byproducts in smaller amounts, the potential negative effect of nutrient variation on cow performance is reduced. When contracting for these commodities, an acceptable range in variation between loads should be agreed upon ahead of time. Loads delivered outside this agreed-upon-range would then be refused.

Wheat Midds

Wheat midds contain 26 per cent starch and 18 per cent protein. Price can be substantially lower than corn at the right time of the year. Wheat midds have not been evaluated as a substitute for corn in many experiments. In a recent study, wheat midds were fed at about 7.5 per cent of the diet, replacing a combination of corn and soybean meal.

Corn was reduced in the diet from 34.1 to 28.9 per cent. Cows fed wheat midds tended to eat less feed dry matter but milk production and milk composition were not affected. This reduced feed intake may have been because wheat midds have a high water-holding capacity, thus increasing gut fill with fibre. Cows fed wheat midds appeared to have looser manure than those fed more corn and soybean meal.

In a second study, wheat midds were fed at 0 or 22.4 per cent of the diet, replacing 35 per cent of the ground corn and 30 per cent of the soybean meal. Intake of feed dry matter, production of milk, and milk composition were not different between the two groups of cows which averaged 150 days in milk at the start of the study. However the digestibility of dry matter and neutral detergent fibre were lower for cows fed wheat midds. Therefore wheat midds do not appear to be an effective feed replacement for ground corn except for lower producing cows.

Dried Distillers Grains Plus Solubles (DDGS)

Dry corn put through the dry-milling process will produce ethanol, carbon dioxide, and DDGS. Each 56-pound bushel of corn processed through the dry milling process results in 2.85 gallons of ethanol, 18 pounds of carbon dioxide, and 18 pounds of DDGS. Plants using the dry milling process are less expensive to build than the plants using the wet milling process, thus about 75 per cent of the ethanol produced from corn comes from dry milling.

The production of DDGS in the US increased tenfold between the years 1980 and 2000, increasing from 320,000 to 3.5 million metric tons (1 metric ton = 2205 pounds). Production doubled again between 2000 and 2004 to over 7.3 million metric tons. Every dairy cow in the U.S. would need to consume 7.3 pounds per day of DDGS over a 305-day lactation in order to use up this supply, but of course DDGS is also fed to other livestock and is exported.

During the production of ethanol from corn, a syrup product and a cake product are produced. These products are blended together in different proportions to form DDGS. Depending on the plant, the DDGS may be a mixture of syrup to cake ranging in proportion from 35:65 to 55:45. Because of the variation in the ratio of syrup to cake used by different plants to form DDGS, nutrient concentrations will vary from one plant to another. However the variation looks to be less than that of CGF. Although DDGS is very low in starch, it is an excellent source of protein (30.3 per cent) and fat (13 per cent) and therefore has an energy density very similar to corn. In addition, the ethanol-making process increases the digestibility of the fibre. Unlike CGF, DDGS is a very good source of ruminally undegradable protein (RUP) but if the temperature of the drying process is too high, the protein may be indigestible.

Purchasers of DDGS should analyse for ADIN regularly. Kalscheur *et al.* reviewed 24 experiments in which 98 comparisons were made between cows fed diets containing DDGS and those not fed DDGS. Cows fed DDGS at up to 30 per cent of the ration produced as much milk as those not fed any DDGS (about 73 lb/day). In spite of decreasing starch content, milk production was maintained. In a study conducted at the University of Florida, whiskey DDGS were fed at 0 per cent or 20 per cent of a 55 per cent concentrate:45 per cent forage diet. The forage was all corn silage in one set of diets and 50 per cent corn silage:50 per cent rhizome peanut silage in another set of diets. The DDGS replaced around 40 per cent of the corn and soybean meal. Milk production was increased from 59.0 to 60.7 pounds per day without changing milk fat test. The effect of DDGS was the same regardless of whether the forage was totally corn silage or an equal mixture of corn silage and rhizome peanut silage. However, a maximum feeding level of DDGS might be 15 per cent in order to prevent the overfeeding of protein, RUP, unsaturated fat, and phosphorus. Increased corn oil in the DDGS can also lead to decreasing milk fat, so fat content of DDGS should be monitored regularly.

Feeding DDGS at 20 per cent of the diet will increase dietary fat by 2.6 per cent. This will likely be a concern if the diet contains other supplemental fats. Dietary fat should be kept below 6 per cent. The feeding of one pound of DDGS can replace about 0.6 pounds of corn and 0.4 pounds of soybean meal.

Soybean Hulls (SBH)

During the processing of soybeans for oil and meal, the hulls are separated and ground. The resulting soybean hulls consist largely of the outer covering of the soybean so they are high in fibre and contain moderate amounts of protein but very little starch. Protein concentration will depend upon how well the hulls are cleaned. Positive characteristics include a high content of lysine, a low content of phosphorus, and a highly digestible fibre. Fat content can vary widely. The availability of SBH is likely to increase in coming years as more acres are planted to soybeans in order to use soybean oil for biodiesel production. In 13 separate experiments published between 1976 and 2002, SBH (fed at up to 20 per cent of the dietary DM) successfully replaced ground corn or high moisture corn in rations for lactating cows. In most of these studies, the dietary starch levels were probably greater than what is recommended today, based upon the proportion of corn in the diets fed in these studies. Therefore the replacement of corn with SBH still left enough starch to support adequate intake and milk production. One study fed a more conservative amount of corn in the control diet and deserves further attention here. The control diet was 23 per cent high moisture corn, 26 per cent corn silage, and 26 per cent alfalfa silage. Adding SBH to the diet at 14 per cent of the dietary DM reduced the corn to 9 per cent of the diet. This dropped the starch from roughly about 25 per cent to 15 per cent, calculated from average values. Diets were fed starting at 8 days fresh.

Lactating cows ate more DM when offered the ration containing SBH but milk production was not changed. Lactating heifers fed SBH ate the same amount of feed and produced the same amount of milk as those not fed SBH. Concentrations of milk fat and protein were unchanged. Replacing 60 per cent of the corn with SBH, so that SBH made up 14 per cent of the ration, was effective to support high milk production in early lactation. Even though dietary NDF increases when SBH replace corn, feed intake has not been decreased. This is in contrast with the well-documented fact that as traditional forage replaces concentrate in the diet, concentration of dietary NDF increases and feed intake decreases. However the NDF in non-forage fibre byproducts like SBH does not have the same physical characteristics as the NDF in traditional, properly chopped forage. The fibre in SBH is highly digestible, very short and has a specific gravity that allows it to move out of the rumen quickly so that SBH fibre does not have the same rumen fill as forage fibre at similar NDF concentrations. In order to use SBH most efficiently, the diet should contain sufficient "effective" fibre from traditional forages.

Corn Silage

Corn silage hybrids differ in the proportion of grain in the total crop. Selecting corn silage hybrids that contain more starch will reduce the amount of ground corn needed. In a test of 55 corn silage hybrids grown in Gainesville in 2006, starch concentration ranged from 22 per cent to 35 per cent of silage dry matter, with an average of 29.3 per cent. Selecting a corn silage hybrid with high levels of starch and digestible fibre, while maintaining high, could be a good strategy to reduce corn costs in the ration. How much could be saved? If a corn silage hybrid had 33 per cent starch instead of 30 per cent, and corn silage was fed at 33 per cent of the ration dry matter, dietary ground corn could be reduced from 18 per cent to 16.5 per cent, which is about 1 lb/ day less ground corn. The cost savings would depend upon the relative price of the feedstuff(s) used. Corn silage contains much more starch than other forages fed in Florida. Feeding more corn silage in the ration and less sorghum silage, cottonseed hulls, or bermudagrass will increase the starch in the diet and allow for less ground corn to be fed. If corn silage inventory allows, increasing the corn silage from 33 to 38 per cent of the diet DM by replacing another forage like sorghum silage, will allow ground corn to be reduced by 1.5 percentage units.

Glycerol

Glycerol is a byproduct made as plant and animal fats are processed to make diesel fuels. Diesel fuel produced from fat burns cleaner, as there is no soot or particulate matter produced. Starting with animal fat to make diesel fuel is more profitable than starting with vegetable oil because animal fats are cheaper. However, diesel made from animal fat may not work as well in colder climates because it "clouds up."

This process is expanding worldwide and therefore the supply of glycerol in the future is likely to increase. Ten pounds of glycerol result from every 100 pounds of biodiesel produced. The purity of this byproduct can vary widely. The more impure the product, the more water, methanol, phosphorus, and potassium it contains. A product with these contaminants is labeled glycerin. The glycerin fed in a German study contained 2.2 per cent potassium and up to 2.4 per cent phosphorus. Methanol is present due to its use in the manufacture of biodiesel.

Although ruminal bacteria can detoxify methanol, it can be problematic if present in large quantities. According to the FDA, glycerin is considered a substance that is GRAS (Generally Recognized as Safe) for general purpose use in animal feed, unless methanol is present at concentrations exceeding 150 ppm. The energy content of glycerol is similar to corn when fed in high starch diets. Dietary glycerol would be converted mainly to propionate and butyrate by ruminal bacteria. As a feed ingredient, it could substitute for corn or molasses.

Little research has been done regarding the feeding of glycerol to dairy cows. Lactating dairy cows fed glycerol (1.3 per cent methanol) at either 0 per cent, 2.7 per cent, or 5.3 per cent of dietary dry matter ate the same amount of feed and produced the same amount of milk. Milk composition was unchanged with the exception that milk urea nitrogen concentrations were lower in cows fed glycerol. Efficiency of fat-corrected milk production was improved from 1.46 to 1.59 and 1.60 lb of FCM yield per lb of feed dry matter intake. Interestingly, animals of other species consumed more water when they were fed glycerol. This may have a benefit to dairy cows managed under heat stress conditions. German researchers fed glycerol up to 10 per cent of ration DM successfully. Glycerol appears to be a good pelleting agent when added at 5 per cent.

New Technologies to improve Ethanol Production

As the ethanol manufacturing industry works at improving the efficiency of conversion of corn to ethanol, different byproducts will become available. Applegate *et al.* lists the following potential corn byproducts.

- Called the "quick germ quick fibre method," an enzyme is added to the water used to soak the ground corn, causing the germ and fibre to float before the fermentation process begins. The product from this process contains 28 per cent protein, 5 per cent fat, and 25 per cent NDF.
- Collecting the pericarp fibre and germ prior to fermentation by modifying the dry grinding process (drum degerminator) results in a product that is 24 per cent protein, 8-9 per cent fat, and 28 per cent NDF.
- Removing the fibre by sieving and air aspiration results in a product that is 40+ per cent protein, 15 per cent fat, and 20 per cent NDF.
- Modifying the yeasts used to ferment the sugar to ethanol could allow for an increased concentration of lysine in DDGS, thus giving DDGS a more favourable amino acid profile.

As each mill adopts what they consider the best ethanol-producing technology, the feeding industry will be faced with a variety of byproducts on the market which will need to be properly identified prior to purchase.

With the increased price for corn due to demand for ethanol, farmers will be planting more of their acres to corn. This will reduce the number of acres committed to other crops, such as cottonseed. This shift will likely have a large impact on the market price of a number of feed commodities. The availability and price of each commodity will need to be evaluated for optimal pricing. Some acceptable, alternative feeds may allow some reduction of corn grain in the diet of lactating dairy cows.

FEEDING GOATS

Goats are browsers, like deer, which means they prefer trees, bushes, and woody weeds; rather than standing still and eating grass down to the roots, they like to stay on the move, eating a bit of this and a bit of that. Goats can learn to graze a pasture, but don't expect it to be "mowed." The grass helps supplement the goats' diet, but low grazing also can spread parasites. Goats have specific nutritional needs, only some of which are met by the plants on your farm that they browse on. You have to provide feed for the needs that can't be met by browsing. Unless you have a lot of property with a variety of browse, feed will be your biggest expense in raising goats. Don't scrimp on goat feed — it will pay dividends in good health, milk production, and lower veterinary bills.

Feeds for goats can be divided into several different categories: forages, grains, protein supplements, minerals and water. Depending on what levels of each category that is fed. It will be found that goats will gain about.25 to.5 a pound each day. Goats will readily eat a variety of forages. These include pasture, hay, haylage, and brush. Any species of forage is acceptable for goats, although be careful of feeding a lot of alfalfa to wethers and billies because the higher calcium levels can lead to urinary calculi or kidney stones. Goats are very useful in many operations for cleaning up unwanted brushy areas. Most brushy areas will only be able to survive about 2 years with intensive defoliation.

Forages are an important part of a ruminant's ration because the high fibre is necessary to keep the digestive system healthy. Grains provide the energy portion of a ration for goats. A variety of grains are available, with corn, oats, wheat and barley being the most widely used. Limit the amount of wheat in the ration to no more than 2 per cent to prevent any digestive problems. When starting goats on a feedlot ration, be sure to gradually introduce them to grain. Feeding high levels of grain suddenly can cause bloating and overeating disease. Disease can be prevented by vaccinating with type C and D perfringens. This is often available in one vaccine, along with tetanus. Any male goats that are castrated should receive a dose of tetanus vaccine. Castration is a personal consideration as many ethnic groups prefer to purchase intact males.

Protein Supplements

Market goats should be fed a grain ration that runs between 14 per cent and 18 per cent protein. To reach these protein levels, we need to include some sort of protein supplement. Typical supplements will include soybean meal, roasted soybeans, cottonseed, brewers or distiller's grains, or flour mill by-products. Beware of using any dairy protein supplements because of the amount of copper in the supplements. Many nutritionists still disagree over the amount of copper required in a goat ration, but care should still be given to overfeeding this nutrient.

Minerals

One of the best ways to provide minerals to a market goat is through a commercially mixed salt and mineral mix. Be sure to purchase a mix formulated specifically for goats. Allow goat's free access to a mineral feeder that contains the mineral mix at all times. Minerals are especially important for helping to ensure that goats remain healthy. These mineral mixes will often include some of the essential vitamins for goats as well.

Water

Water is often one of the most overlooked aspects of livestock feeding operations. Clean, fresh water should be available at all times for all species of livestock. Animals that have adequate amounts of water are more likely to stay healthy as well as have a faster growth rate. Goats will often be able to supply all of their water needs through dew and lush pasture. However, it is still in their best interest to have a supply of water available to them at all times.

FEEDING YOUR GOAT HERD

Feed is the largest cost associated with raising goats. It can affect herd reproduction, milk production and kid growth. Late gestation and lactation are critical periods for doe nutrition. Nutrition level determines kids growth rates. Goats not receiving adequate nutrition are more prone to disease and will fail to reach their full potential. Goats require energy, protein, vitamins, fibre and water. Energy is the most limiting nutrient, while protein is the most expensive. Imbalances of vitamins and minerals can limit animal performance and lead to various health problems. Fibre maintains a healthy rumen enviroment and prevents digestive disturbances. Inadequate water intake can cause various health problems.

Water: Water is the cheapest feed ingredient and often the most neglected. Goats should have free access to clean fresh water at all times. It is critical that they have an adequate supply of water in the Winter months. A mature goat will consume between 3/4 to 1 1/2 gallons of water per day.

Pasture and Browse: Pasture and browse are the most economical source of nutrients for goats. Pasture tends to be high in energy and protein in its vegetative state. As the pasture plants mature the palatability and digestibility decline, making it important to rotate pastures to keep plants in a vegetative state. Some of the best pastures for goats are Bahiagrass, millet, sorghum, sudan grass and a mixture of a grain, grass and clover. During the early part of grazing season, browse tends to be higher in protein than ordinary pasture. Goats are natural browsers and select plants at their most nutritious state. Goats that browse have less problems with internal parasites.

Hay: Hay is the primary source of nutrients for goats during the winter months. Hay varies in quality and the only way to know the nutritional content is to have it analyzed by a forage testing laboratory. Legume hays- alfalfa, clover, lespedeza- tend to be higher in protein; this also depends on which cutting it is.

Briers and such: Goats love eating multifloral rose bushes, green briers, poison ivy and just about every type of unwanted brush located on your property. Many herds are used specifally to "clean" up overgrown property, which can be more economical than hiring people to do the job.

Vitamins and Minerals: Many mierals are required by goats. The most important are salt, calcium and phosphorus. Vitamins are needed in small amounts. Goats require vitamins A, D, and E. Offer a pre-mix of loose minerals free choice; goats will consume more if loose is available.

Grain: It is often necessary to feed grains when forage alone cannot provide enough nutrients. CREEP FEEDING and supplemental feeding of kids does increase growth weight, but should only be done to the extent that increases profits. There are two types of feed- carbon and protein. Carbon or 'energy' feeds include cereal grains- corn, barley, wheat, oats, milo, and rye- various by products feeds such as fat, soybean hulls and wheat middlings. Protein supplements may be of animal or plant origin and include soybean meal, cottonseed meal and fish meal. 14% CP to 16% CP will usually fill the nutritional requirement for various classes of goats.

OTHER FEED ADDITIVES

For herds that are prone to urinary calculi, ammonium chloride can be added to prevent the disease. Another common practice is to add a coccidiostat to the ration to prevent as well as control coccidiosis. Common additives used to prevent coccidiosis include Rumensin, Bovatec or Deccox (decoquinate). Lastly, no matter what the ration, be sure to provide enough feeder space for all the goats to eat. Plan for at least 1 foot of feeder space for each goat. And, don't forget to clean feeders and waterers on a regular basis. Sound feeding practices and good housing facilities result in optimum growth and high milk production and contribute to the good health and comfort of goats.

When feeding goats, keep these objectives must be in mind:

- Feed a young animal enough energy for growth, and feed a mature animal enough energy to maintain a fairly constant body weight;
- Provide enough protein, minerals and vitamins in a balanced feeding programme to maintain a healthy animal; and
- Offer does enough extra food during gestation and lactation for fetus development and milk production.

Optimum growth, good health and high milk production are the results of sound feeding practices. Dairy goats are not unique in their body requirements; they will respond to good nutritional practices. Digestible fibre is especially important in dairy goat diets. Too much grain in relation to forage does not foster good ruminant action and is a costly feeding practice. When feeding, keep minerals and trace mineral salt separate. Always feed them separately.

Feed hay in a rack that will not permit wasting. One recommended type of feeder is a keyhole feeder. Do not overlook forage testing as an economical way to feed the correct ration to herd.

Feeding Kids and Yearlings

Kids are born with no natural protection from disease. The first milk (colostrum) from the mother offers protection and gets the digestive system working. To be most effective, it must be fed before disease-producing organisms enter the mouth and digestive tract. Wash the fresh doe's udders and teats with warm water.

Hand milk half a cup of colostrum and feed it to the kid within 15 minutes of birth. This is the best way to ensure that the newborn receives some milk and to provide it with the most protection from organisms present on the skin of the doe. Complete the first milking and store the colostrum for later feedings if hand feed is elected. Otherwise, permit the kid to nurse at its convenience following the first hand feeding. Clean the feeding utensils immediately after each use.

The same cleaning procedures should be followed for washing milk-handling equipment. Milk replacer may be fed from the fourth day; it should contain at least 20 per cent protein, 20 per cent fat and be free of vegetable products. A lamb or high-quality calf milk replacer is recommended. Provide hay and grain at one to two weeks of age. Wean from milk when grain intake reaches 1/4-pound daily and kids are readily consuming hay. Make sure the solution is thoroughly mixed. Mix a fresh mixture daily and feed in place of milk. Feed as soon as diarrhea is noticed. Use for 1-1/2 or 2 days, then return to the regular milk diet. After four to six months of age, the kids may be fed a ration similar to that of the milking herd. Good hay and 1/2 pound of grain per day should provide an ample growth rate. Poor hay may require 1 to 1-1/2 pounds of grain daily.

Feeding the Milking Herd

If milk production is important, feed maximum amounts of high quality hay balanced with a grain ration containing enough protein, minerals and vitamins to support production and animal health. Grass or legume hays are equally acceptable.

As the percentage of legumes is increased, the need for protein in the grain mix is reduced. To determine the amount of grain to feed, consider level of milk production, amount and quality of forages consumed, appetite and state of fleshing. Thin, high-producing does should have access to all the hay they can eat plus grain to the limit of their appetite. Does in mid-lactation that are in good flesh should have all the hay they will eat plus 1 pound of grain for each 3 pounds milk produced. Late lactation does may not need more than 1 pound of grain for each 5 pounds of milk. Feed a grain ration formulated for a milk-producing ruminant (dairy cows). Rolled or cracked grain is more palatable than ground grain. Because of palatability problems, urea is not recommended. Some commercial cow feeds may contain byproduct ingredients unpalatable to goats.

Wet molasses is more palatable than dry molasses. Beer or citrus pulp is a valuable source of fibre, especially if the available hay is of low quality. If the doe is not thin, reduce the amount of grain to 1/2 to 1 pound per day. Feed her all the forage she will eat.

Hay fed during the dry period may be of lower quality, but if so, the grain ration should contain additional protein. Browse, leaves and weeds are often useful to recondition the stomach. If the dry ration differs from the milking ration, be sure to change to the milking forage and grain ration two weeks before the doe freshens. Most bucks do not need more than a pound of grain per day plus forages. Don't let them grow fat. Adjust grain upward or downward accordingly. Always feed full forage.

SHEEP DIET

Mostly sheep eat grass, clover, forbs, and other pasture plants. They especially love forbs. It is usually their first choice of food in a pasture. A forb is a broad-leaf plant other than grass. It is a flowering plant. Forbs are often very nutritious. As compared to cattle, sheep eat a greater variety of plants and select a more nutritious diet, but less so than goats. Sheep are exclusively herbivorous mammals. Like all ruminants, sheep have a complex digestive system composed of four chambers that allows them to break down otherwise inedible cellulose from stems, leaves, and seed hulls in to complex carbohydrates. When sheep graze, vegetation is chewed in to a mass called a bolus, which is then passed in to the first chamber: the rumen. The rumen is a 5-10 gallon (18.9-37.8 litre) organ which ferments feed via a symbiotic relationship with the millions of bacterium, protozoa, and yeasts which are collectively called flora.

The bolus is periodically regurgitated back in to the mouth as cud for additional chewing and salivation, this process usually takes around six hours. Cud chewing is an evolutionary adaptation that allows ruminants to graze more quickly in the morning, and then fully chew and digest feed later in the day. This is beneficial because grazing, which requires lowering the head, leaves sheep more vulnerable to predators, while cud chewing does not. During fermentation, the rumen produces excess gas which must be expelled; disturbances of the organ, such as sudden changes in a sheep's diet, can cause potentially fatal conditions such as bloat. After fermentation in the rumen, feed passes in to the reticulum and the omasum; special feeds such as grains may bypass the rumen altogether.

Following the first three chambers, food moves in to the abomasum for final digestion before processing by the intestines. The abomasum is the only one of the three chambers directly equatable with the usual mammalian stomach, and is accordingly sometimes called the "true stomach". Sheep follow a diurnal pattern of activity, beginning to feed at dawn and throughout the day until dusk, stopping sporadically to rest and chew their cud. Ideal pasture for sheep is not lawn-like grass, but an array of grasses, legumes and forbs. Types of land where sheep are raised vary widely, from pastures that are seeded and improved intentionally to rough, native lands. Common plants toxic to sheep are present in most of the world, and include (but are not limited to) oak and acorns, tomato, yew, rhubarb, potato, and rhododendron. Sheep also may perish from consuming man-made substances.

Sheep are largely grazing herbivores, unlike browsing animals such as goats and deer that prefer taller foliage. With a much narrower face, they can crop plants very close to the ground and can overgraze a pasture much faster than cattle. For this reason, many shepherds utilize some type of managed grazing, whereby a flock is rotated through multiple pastures, giving plants time to recover. Paradoxically, sheep can be both the cause of and solution to one of the major ill effects of overgrazing: the spread of invasive plant species. By disturbing the natural make-up of pasture, sheep and other livestock can pave the way for invasive plants. But sheep also prefer to eat invasives such as cheatgrass, leafy spurge, kudzu and spotted knapweed over native species such as sagebrush—thereby making grazing sheep an effective measure in restoring native pastures.

Other than forage, the only other common staple feed for sheep is hay, most often during the winter months. The ability to survive solely on pasture (even without hay) varies with breed, but all sheep are capable of doing so. Also included in most sheep's diets are minerals, either in a trace mix or in licks. Naturally, a constant source of potable water is also a fundamental requirement for sheep. The amount of water needed by sheep fluctuates with the season and the type and quality of the food they consume. When sheep feed on large amounts of new growth and there is precipitation (including dew, as sheep are dawn feeders), sheep need less water. When sheep are confined or are eating large amounts of cured hay, more water is typically needed. Sheep also require clean water, and may refuse to drink water that is covered in scum or algae.

COOLING PONDS FOR DAIRY CATTLE

Cattle have used water for cooling probably for as long as there have been cattle. The last survey done in Florida reported that 30 per cent of the dairies had cooling ponds. The reader should be aware that cooling ponds, if not maintained properly, could harbour infectious diseases such as Leptospirosis and many mastitis organisms. The water usage is very high to fill and keep full.

Environmental concerns that must be addressed are the run-off from ponds and ground seepage. Every state seems to have many regulations that must be addressed before building cooling ponds. The benefits of cooling ponds and some suggestions on building and maintaining them. It is the dairymen's responsibility to use them properly and to be aware that ponds can harbour diseases and the pond water must be dealt with according to state and local laws.

What is a cooling pond for dairy cattle? For this discussion they are a man made hole in the ground with a constant inflow of clean fresh water. Usually the ponds drain or overflow from the top. If a series of ponds is used, the overflow water may be run into a retention pond to be irrigated on crops or enter the waste management system. Cooling ponds may also be fed from springs or artesian wells. So cooling ponds for dairy cattle should have constant fresh water in and a constant outflow of water to some place for proper disposal. Cement ponds may also be built. These may drain from the bottom and can be used to flush feed lanes or barns. Bottom draining will help remove some of the solids that have settled to the bottom of the pond. Care must be taken on the entrance and exit to cement ponds. These may be very slippery if the cement is not cross-grooved. A stagnant pond without fresh water entering, a lagoon, a wallow, and a mud hole do not classify as a cooling pond for dairy cattle.

Site Selection

A site near the feed and drinking water, ideally with some form of shade for cows, is best. The idea is for cows to cool, leave the pond and goes eat and drink, lay down in the shade and then do it all over again. If cows do not have feed and drinking water near, they will stay in the pond and not eat and we have cool wrinkled cows that do not give much milk.

Cooling ponds should be located close to a water supply as fresh water needs to enter the ponds constantly. There must be a way to dispose of runoff water properly. This will vary by areas of the country. Some slope from the ponds to the waste management area would be helpful, so pumps need not to be used. Cooling ponds should not be too far from the milking parlor as long walks in the sun increase heat stress. Exit lane sprinklers help make the long walk more comfortable.

Concrete Ponds

Concrete ponds will work and have been used but they are expensive to build. However, they are easier to maintain, as the entrances and exits do not have to be rebuilt. The finish of the concrete on exit and entrance slopes should be cross-grooved to prevent cows from falling. A 1:8 slope would be a safe slope if concrete ponds are constructed.

Sizing Cooling Ponds

There is no official listing for cooling pond size. If a holding area requires 15/ft^2/ cow and cows are driven into a pond and taken out as a group then 15/ ft^2/ cow might be adequate. If 50/ft^2/ cow is recommended for most general animal space, then this might be the correct figure for a cooling pond. This amount of space would allow cows to enter and exit the pond at all times when all cows were not using the pond at the same time. If 50/ft^2/ cow for a pond is used, then 5000/ft^2 is then needed for the 100 cow group.

Pond Shapes and Depths

Using the sample of 5000/ft^2/100 cows a pond could be 50' wide × 100' long, or a circle 80' in diameter (A= (Pi) r^2; A= 3.14 × 40 = 5024ft^2). If a 50' × 100' pond is used and the sides are fenced so the cows can only enter or exit at each end of the pond, the amount of dirt falling into the pond will be reduced. If a circular pond is made, cow carpet should be used to eliminate the problem of dirt falling in the pond. As cows can swim, pond depth is not a problem. Deeper ponds may allow the settling of organic matter on the bottom of the pond. This organic matter may not be disturbed by cows walking through it as they will swim over the top of it. Preference should be given to depths of 3' or 4' as it seems to cool the cows and let them stand and walk in and out at will. While deep ponds allow cows to submerge, they do not seem to like to float too long so they go to where they can stand and may block cow traffic. A circular pond that is deep in the middle lets cows dunk themselves while others enter and exit the pond at will.

Water Usage

A 100 cow group with 50/ft^2/ cow could have a 50' wide × 100' long × 3' deep cooling pond (it may be 4' in the middle and slope in and out). A cooling pond 50' wide × 100' long × an average 3' deep=15000/ft^3. There are 7.48 gallons/ ft^3, thus the pond requires 112,200 gallons of water.

The pond must be filled and then kept fresh with running water at all times while in use. A 1" pipe 300 ft. long and water pressure at 35psi will result in a flow rate of about 23.5 gpm. At this rate it would then take 80 hours to fill the pond initially. The dairy water usage would then be 1,410 gph × 24 hours or 33,840 gallons of water per day. This fresh water must be pumped into the pond and there is going to be close to that amount of run off water that will exit the pond for disposal.

Water Run-off

Disposal of run-off water must be in accordance with whatever local, state, and federal laws that applies. Many dairies run the water into the waste management system, while others add it to retention ponds to be used for irrigation.

Shading over Ponds

Experience has led to discontinued use of shade over ponds as the cows never leave the ponds. Remember, the object of this exercise is to increase dry matter intake and cows will not eat much soaking in the pond.

Pond Maintenance

At least once a year, usually during the winter, the pond should be fenced and the water removed by pumping. The mud and manure on the bottom of the pond should be removed and the pond left empty for the sun to shine on it. In the spring, the pond should be filled again and the fences removed. Some people just fence the pond in the winter, dig a new pond and fill in the old one with new pond's dirt.

Mycoplasma mastitis is often cultured in water from cooling ponds. This is a real problem if pulsators do not work well resulting in damaged teat ends.

Many herds with mycoplasma positive ponds, have no mycoplasma in the bulk tank. If the water is filthy, cows should not be permitted in the ponds and the ponds should be pumped dry. The sludge should be cleaned and the ponds filled with water again. Both the sludge and the water shoud be removed from cow areas.

Cooling Ponds to Reduce Body Temperature

In the summer of 1986, an experiment was conducted at a North Florida dairy that was using man-made cooling ponds. Ten early lactation Holstein cows were fitted with radio transmitters that transmitted inner ear temperatures every five minutes to a radio receiver and data logger.

Data were collected over a four-day period in mid-August. The cows' activities were also monitored during this time. Entering and exiting the ponds, eating, drinking and laying were recorded.

The cooling ponds lowered the cows' temperature by 1-2°F depending on the time of day they entered the cooling ponds. The average length per stay in the pond was 18 minutes for events from midnight to noon and 12 minutes per visit from noon to midnight. It was obvious that cooling ponds cooled cows.

Pond Water Quality

In 1987, samples were taken from a dairy's man-made earthen ponds. The ponds were sampled weekly from May 19,1987 to September 28, 1987. Six to ten ponds were sampled each week.

Total bacteria counts (TBC), varied greatly from week to week. This may have been due to no water entering the ponds. There did not seem to be any great increase in total bacteria counts or coliform counts as time progressed. The total bacteria count averaged 3,133,700 CFU/ml for all ponds for the 20 weeks. The coliform counts averaged 14,340 CFU/ml for the same period.

Cooling Ponds and Their Effect on Milk Quality

In a previous study on a dairy in Florida with man-made cooling ponds, it was reported that cows exposed to cooling ponds did not experience more clinical mastitis than cows that did not have access to cooling ponds. In fact, cows exposed to cooling ponds during the trial period were only half as likely to develop a case of clinical mastitis as cows not exposed to cooling ponds.

Clinical Mastitis Organisms

The same dairy had provided its clinical mastitis records for several years. The total number of clinical mastitis cases was high in the first quarter of the year and then declined sharply about the time hot weather came and the cows started using the ponds. The incidence stayed low, even in the last quarter of the year when the ponds were not in use. This can be explained many ways. The first quarter of the year was an extremely wet period.

The cows were quite dirty and the cow wash system was available to one-half of the herd. This may explain some of the variation, even though one-half of the herd had never had a cow wash. Another confounding variable was that this herd switched to Clorox for pre- post-teat dipping in April of that year. It had previously pre- and post-dipped with a Chlorhexidine product. From this it could be concluded that Clorox caused the reduction of clinical mastitis. However, there were no controls in this study, one could conclude nothing except that the ponds did not increase clinical mastitis.

This herd was on a lactation treatment study during the summer of 1987. A total of 40 cows were sampled and treated. There were no unusual organisms treated. These 40 cows were not all the cows treated for clinical mastitis during the period the cows had access to the ponds, but the results were encouraging.

Statewide Effects of Ponds on Milk Quality

Regulatory samples for all the herds in Florida were obtained from the Division of Dairy Industry for the year of 1987 and were analysed for the effects of no ponds, man-made ponds and natural ponds. The numbers presented are least squares estimates of Standard Plate Counts (SPC) and Direct Microscope Somatic Cell Counts (DMSCC) for each month. The data are not biased by different numbers of dairies with no ponds, natural, or man-made ponds. The data were analysed by herds whose milking cows had access to ponds as indicated on the survey.

The data indicated that milking cows with access to man-made cooling ponds had lower SPC and DMSCC counts than other herds in the state with no access to ponds and those herds with natural ponds were higher in both categories. From these data it could be argued that people who built man-made ponds were better managers and thus had higher milk quality.

FEEDING AND NUTRITION OF PIG

Feeding guinea pig is a fairly simple manner, though there are a few special needs that we should be aware of. Guinea pigs are strict vegetarians, and in addition to their pellets, they will eat fruits and vegetables, bread, and will even nibble on vegetarian dog biscuits. Before getting in to the details, it is important to point out that guinea pigs like a routine, predictable life. Sudden changes in their diet can cause stress, so be careful. If the brands of pellets is changed, gradually by adding a few of the new ones to their old pellets, and slowly switching the mixture over. Feeding excessive amounts of fruits and vegetables at one time can cause diarrhea, particularly if new foods are introduced.

All guinea pigs need water to survive. Though they can obtain a large portion of their water from fresh fruits, vegetables and greens, they will still require water from a water bottle to stay healthy. This water should be changed daily, and the water bottles themselves washed periodically to keep them clean. To prevent algae growth, make sure that the sun does not directly shine on the water bottle for long periods of time. If there is arrangement of tap water bottled water or a water filter system should be considered. Distilled water is not recommended for regular drinking, as it can disturb the osmotic pressure in the body cells, which can potentially lead to osmotic shock.

Be aware that some guinea pigs like to "play" with their water bottle, leaking out large amounts of water by holding the steel ball up in the tube. Others like to try and blow water back up into the bottle, which can contaminate the entire supply. Hay is the mainstay of every guinea pig's diet. Though guinea pig pellets do contain hay, it is not in sufficient quantity to keep guinea pig healthy. Guinea pig must have a fresh supply of hay every day in order to keep its digestive system regular, and as a general rule.

Timothy hay is the best type of hay to feed, though any grass hay will do. If grass hay is not available, a legume hay can be fed, such as alfalfa, though in general this should be avoided as much as possible. Legume hays are high in calcium, and this can lead to bladder stones (uroliths) in some animals. Most pet stores carry dried timothy hay in large and small bags, but it is far better to purchase hay fresh from a feed and garden store, or directly from a farmer. Buying hay by the bale is especially economical, as a bale of hay will literally last for months, and will stay fresh as long as it is stored properly. Some tips on hay storage can be found here.

Guinea Pig pellets

Guinea pig pellets will provide guinea pig with the proper balance of vitamins (save for Vitamin C), minerals and other nutrients. Although it's not necessary to feed pellets to guinea pigs, it is certainly recommended that they be primary feed, after hay. Guinea pig pellets are, in particular, a prime source of protein; obtaining the necessary amounts of this and other nutrients will require careful dietary planning if pellets are not chosen. As a general rule of thumb, adult guinea pigs will eat between one and two ounces of pellets each day. More active animals will eat less, and animals without much stimulation will eat more out of boredom. Nursing and pregnant sows may demand a little more than this, and animals that are given plenty of hay and some fresh vegetables each day might eat a little less. It will be easy to tell if they are fed too much or too little: if there are leftover pellets. If the bowl is empty and they seem to be foraging for food, then they aren't fed enough.

Try to feed at the same hour each day, as guinea pigs like a routine and predictable life. Remove any leftover pellets after an hour or so. Guinea pig pellets are a prime source of protein: a diet that is heavy on pellets will lead to a fat guinea pig. Hay is the only food that they should have access to 24 hours a day. When choosing pellets, be sure to *avoid* the guinea pig "mixes" that contain nuts, seeds and dried fruits. These mixes are high in fats and oils, which can lead to excessive weight gain. Additionally, many of these mixes contain sunflower seeds in their shells. Guinea pigs should *never* be fed nuts or seeds that are still in their shells (peanuts, sunflower seeds, etc.): dozens of guinea pigs in the Pacific Northwest alone die each year from choking on these shell fragments.

One thing that should be noted before we move on: guinea pigs should be fed guinea pig pellets, not rabbit pellets or pellets for another species of animal. Guinea pig pellets are nutritionally balanced for guinea pigs, and feeding them pellets for other animals can result in serious health problems. Though some people may tell that their guinea pigs survive well on rabbit pellets, for instance, it is not trusted that all rabbit food is safe for guinea pigs. Their particular brand may work for guinea pigs, but another brand may lack necessary vitamins or minerals, or have them in insufficient or excessive quantities.

There is no regulation in the pet food industry, so pellets and feeds vary considerably between manufacturers. Don't play Russian roulette with cavies: only feed them pellets that are explicitly and *exclusively* for guinea pigs. It is *extremely important* that guinea pigs receive enough vitamin C each day to prevent scurvy. Guinea pigs cannot manufacture or store vitamin C, so they must obtain it from their diet on a daily basis. Adult guinea pigs will require 10mg of vitamin C each day, and nursing and pregnant sows will require twice that amount (20mg). The caveat: vitamin C, also known as ascorbic acid, breaks down very quickly, so it must be supplied fresh. Although many pellet manufacturers will claim that their feed contains the required amount of vitamin C.

Vitamin C in pellets breaks down during storage, and after about 90 days, there won't be enough vitamin C in the feed to keep guinea pigs healthy. Also note that the food starts breaking down on the shelf, not when the bag of pellets are opened.

Unless the manufacturer dates the bag, showing when the feed was packaged, we really have no idea how long pellets have been on the shelf at the local pet store. Since pellets are generally an inadequate source of vitamin C, most owners will either feed sufficient quantities of fruits and vegetables rich in vitamin C each day, or provide vitamin C in the form of supplements added to the guinea pigs' food or drinking water.

Vitamin supplements come in many forms: pre-packaged vitamin supplements for guinea pigs can be bought from pet store, or self grind up vitamin C tablets. In either case, most people prefer to add the vitamins to the animal's drinking water, as that is the only way to guarantee that it will be ingested daily. If the supplement routes are chosen, here are a few pointers:

- Add the appropriate amount of supplement to the amount of water that the guinea pig will drink in a day. Most water bottles work best if they are filled halfway or more, so end up wasting a lot of vitamins. Fortunately, they aren't very expensive. If guinea pig drinks 4 fluid ounces of water a day, and fill water bottle with 8 ounces, then need to add 20mg of vitamin C supplement. This will lead to 4 ounces of wasted water and 10 mg of wasted vitamin C.

- Vitamin C (ascorbic acid) breaks down very quickly in water, so need to add it on a daily basis.
- The chlorine in tap water can actually inactivate ascorbic acid. If vitamin supplements in water is opted, then use tap water that has been standing for at least 24 hours, or filtered tap water from a filter system that reduces chlorine content.

Being vegetarians, guinea pigs will eat many kinds of fruits, vegetables and fresh greens. And, many fruits and vegetables are sources of vitamin C, providing a natural way to meet guinea pig's daily requirement. Calcium and Phosphorus totals for the given food amounts are presented as well. Although these are necessary minerals, and a part of every guinea pig's diet, they are getting them from their pellets, as well.

The nutrients guinea pig is getting along with their vitamin C. Too much calcium can cause bladder sludge, a precursor to bladder stones. The proper balance of phosphorus, calcium and vitamin C prevents guinea pigs from needing vitamin D supplements.

Although most guinea pig pellets contain vitamin D as a precaution, there's no guarantee that particular brand of feed does. In addition to providing vitamin C and variance in cavy's diet, fresh greens and fruits contain a significant amount of water. Feeding "wet foods" will aid nursing sows in lactating, and will reduce the amount of water that the average guinea pig will drink from the water bottle. It is especially useful to take lettuce, green peppers and other watery greens on trips to provide moisture for guinea pigs while in the car, since water bottles will leak heavily as the bumps in the road rattle the ball in the drinking tube.

To reduce or completely eliminate pellets from guinea pig's diet, they can survive on water, hay and fresh fruits and vegetables. However, this will require very careful dietary planning on part, to make sure that they receive the necessary nutrients in the proper amounts. Also, be aware that many fresh greens are laxative in action, which means that run the risk of giving guinea pigs loose bowels, or even diarrhea, if the are fed too much at one time.

If runny droppings are noticed immediately cut fresh greens out of their diet and feed dry foods until the feces returns to normal. Vegetables that are not laxative (such as carrots) may still be fed. Guinea pigs will not eat what they don't like: some cavies have very discriminating tastes, while others will eat anything that put in front of them. Each pig has its own preferences, and it may take some time to figure out what he or she likes best. Given below is a list of foods that guinea pigs can/will eat. This list is neither comprehensive nor complete. Apples, bananas, bread (slightly stale & crunchy, but not moldy), broccoli, carrot greens, carrots and baby carrots, celery (cut into small pieces first), cilantro, cucumber, dandelion greens, grass, green & red bell peppers, green leaf & romaine lettuce, kale, kiwi, mustard greens, oats, oranges, parsley, raspberries, spinach, tomatoes. When feeding "wild" greens, such as grass and dandelion greens, make sure they have not been sprayed with chemicals, or contaminated by droppings or urine from other animals, such as cats, dogs and birds. The dangers of pesticides are obvious, and feces can carry any number of parasites which can be transmitted to cavies eating contaminated greens. Even if dogs and cats are dewormed regularly, their feces can contain protozoa/bacteria which can cause debilitating diarrhea in a guinea pig.

Some foods to avoid are listed below:

- Long celery stalks
- Iceberg lettuce
- Any shelled nuts or seeds
- Raw beans
- Rhubarb

There is some confusion as to whether or not potato peelings are good or bad: some books indicate that they are poisonous to guinea pigs, while others say that they are okay in small amounts. The truth is that potatoes are *okay*, however, any *green* in a potato *is poisoinous*. If potato peelings are given, make sure there are *no green spots* anywhere in the portions. Also, some people may recommend yogurt in small amounts, for it's bacteria-growing properties (to aid the digestive system). Instead, look into acidophilus powders or liquids: acidophilus is also a bacteria growing culture, and it';s likely to give better results since less of it needs to be fed.

There are several commercially available treats on the market that are aimed at guinea pigs. Berry flavoured "crunchies" and flavoured chew sticks are popular among some owners and guinea pigs, and are generally safe. Some owners also have experimented with vegetarian dog biscuits and dry cereals such as Cheerio's.

However, the guinea pig "treat sticks", which are also available commercially. These sticks are essentially seeds and nuts that are held together with honey. Most of the treat sticks on the market contain sunflower seeds that are still in their shells, which is a big no-no for guinea pigs due to the choking hazards.

FISH FEEDING

Although fresh animal by-products such as liver and blood are available in limited quantities and could be used by a small-scale trout farmer (< 5 tonnes fish/year) within an in-house produced moist diet, a dry diet formulation and feeding regime is recommended for the Project fish farms.

Three dietary formulations are recommended for use within three distinct feed lines, namely starter, fingerling and production diets. Table shows the recommended nutrient levels and dietary formulations for the three feed lines, respectively. The formulations are conservative in that emphasis has been placed on using high dietary inclusion levels of known quality ingredients (ie. such as fishmeal; Table) rather than experimenting with unknown cheaper alternatives such as rapeseed meal. For example, although oilseeds (ie. soybean meal, rapeseed meal) and animal by-product meals (ie. meat and bone meal, blood meal) can and have been successfully used at high dietary inclusion levels within practical salmonid rations, their nutritional success is dependent upon the individual manufacturing process used to produce them.

Thus, since the level of technology available in Qinghai Province to manufacture these ingredient sources is still in its infancy, and the proximate composition of these ingredient sources is highly variable, their inclusion level within the proposed diet lines has been kept to a minimum.

Providing fish with the right type of food in the right proportions is a very important task. This can be much more of a challenge in a saltwater tank than in a freshwater tank, because most marine fish are still wild-caught, and their natural diet can be difficult to maintain in captivity. With the right information, however, most marine fish will learn to eagerly anticipate their feeding time and will voraciously devour their food.

Fig. Fish Food

A good rule of thumb to follow is that fish should be fed no more than they can consume within a five minute period. Any food that is left behind after feeding time should be removed before it has a chance to begin decomposing. If it is found that fish are consuming the food in far shorter a time than five minutes and seem anxious for more, the amount that are fed must be increased.

On the other hand, if fish are healthy, but they are only consuming about half of what they are fed, probably need to start feeding less. Always keep in mind that much more harm can be done to fish by overfeeding than by underfeeding. In the wild, reef fishes will browse for food throughout the day.

Their natural habitat as closely as possible, it is recommended that fish must be fed several small meals every day. At the very least, feed two medium-sized meals per day, one in the morning and one at night.

Types of Fish Food

The kind of food to feed depends on the kind of fish are feeding: herbivore, carnivore, or omnivore.

Herbivores

It is common for fish that graze on algae in the wild to suffer in captivity because they don't receive the appropriate type of food to keep them healthy. Many of these fish are fed terrestrial greens like spinach and romaine lettuce, but these are not appropriate foods for these fish that are solely algae eaters. They should be fed either dried or fresh algae. Dried algae is sold in paper-thin sheets that can be broken down into appropriate portion sizes, depending on the number of fish are feeding.

Fresh algae can be obtained by placing a glass container, with one rock in aquarium, on a sunny windowsill. Wait for the rock to grow a nice layer of algae, and then place it back in the aquarium for fish to graze on. Herbivorous fish can also be fed one of the frozen fish foods that is formulated especially for herbivores.

This food should contain a variety of marine algae, vegetable matter, and seafood. Whether herbivorous fish are fed dried algae, fresh algae, or a frozen food formula, herbivorous eaters are grazers and should have food available to them at all times, unlike most other fish.

Carnivores

Meat-eating marine fish can enjoy a variety of fresh, frozen, freeze-dried, and live foods. Fresh foods suitable for fish include many seafood items, such as shrimp, scallops, mussels, clams, and some fish. If seafood is fed to fish, cut it into bite-sized pieces (or serve it whole to larger fish). Fresh seafood can be frozen and then thawed as needed. Frozen foods are available from most fish shops and other sources.

These offerings may include brine shrimp, bloodworms, and daphnia. They may be in the form of a slab that can be broken into smaller pieces or small, ready-to-go cubes. To feed frozen foods to carnivorous fish, thaw a small portion in aquarium water, and then drop it in the tank. Several freeze-dried foods for carnivorous fish are readily available, such as plankton, krill, brine shrimp, and bloodworms. Freeze-dried foods are foods that have been completely dehydrated, yet they maintain most of their nutritional value. They have a relatively long shelf life, particularly when compared to fresh and frozen foods.

Freeze-dried foods should be used as supplements for carnivorous fish. Live foods are almost always a favourite of fish. Adult brine shrimp, bloodworms, and Mysis shrimp are among the better choices. Earthworms are also a great live food choice because they come in all sizes, fish love them. Be sure to thoroughly rinse them before feeding them to fish.

Omnivores

The majority of marine fish are omnivorous, which means they need to eat both meat- and plant-based foods. One easy option for omnivorous eaters is commercial fish food, such as flakes or pellets. However, offering a varied diet will give healthier, more colourful fish. Try to feed fish two feedings a day of commercial fish food and one feeding of a meaty item, such as bloodworms or chopped fish. The food offered must be alternated in addition to commercial food to keep the fish from getting bored and to make sure all of their nutritional needs are met.

9

Modern Techniques for Animal Feed Analysis

Many rapid analytical tests have been developed over the last few years and are of use when faced with a need for quick decisions or when confronted by large numbers of samples, *e.g.* at entry points of shipments, trading situations, on contamination sites or in plant breeding programmes. This review will cover the main issues which prompted developments in quality assurance and control of analysis.

Information will also be given on accreditation of laboratories, together with useful addresses. Subsequent sections will describe standard and widely accepted methods, highlight areas that require particular attention and refer to recent developments in feed analysis. Topics covered will include: sample preparation, analysis of major components (dry matter, ash and minerals, crude protein, fat, fibres and starch) and of secondary plant products (tannins, mycotoxins and other contaminants). Developments in the analysis of whole samples by near infrared reflectance spectroscopy will be mentioned and the potential of this technique to by-pass traditional feed analysis by directly predicting animal responses.

QUALITY CONTROL, ASSURANCE, ACCREDITATION AND PROFICIENCY TESTING IN FEED ANALYSIS

The need for agreement on methods to obtain comparable, useful data became obvious with the beginning of 'scientific agriculture' in the 19th century (Midkiff, 1984). Different methods led to widely varying results and to regulatory confusion. In 1884, the first meeting of the Association of Official Agricultural Chemists (AOAC) began to tackle fertilizer analysis and soon, 1886, included feedstuff testing.

The AOAC president stated in 1896 'The matter of the analysis of foods and feedstuffs, as shown by the experience of the association, is one of the most difficult questions connected with the work of this organisation'. The AOAC was renamed in 1965 to 'Association of Official Analytical Chemists' to take cognizance of the fact that the AOAC now had a wider scope (Midkiff,

1984). A report in 1985 on the quality of data relating to Pb and Cd analysis in food laboratories concluded that results were inaccurate and the validity of the data was in doubt (Patey, 1996). Similar conclusions have been reached for a wide range of data relating to feed analysis. It would appear that feed analysis is somewhat lagging behind food analysis. Selected examples are given below, in order to illustrate the magnitude of the problem.

Bailey and Henderson (1990) concluded that there was an urgent need to improve oil and sugar determinations, since these methods had relatively poor precision amongst 15 feed laboratories. These data are commercially important as they are used when making labelling declarations and for energy content estimation. Lanari *et al* (1991) commissioned a study of brewer's grain, dried beet pulp, lucerne and hay with 20 feed analysis laboratories. Unacceptably large coefficients of variation were found for mean oil content (determined as ether extract): 3.5% (cv = 17.8%); lignin: 5.4% (cv = 27.3); NDF 55.6% (cv = 7.4%). Coefficients of variation were 5% or less for DM, CP, CF, ash, ADF, and GE.

Beever *et al.* (1996) submitted two contrasting maize silages to 10 commercial feed laboratories. They concluded that current feedstuff analysis provided unacceptable variation and required national standards, as coefficients of variation were12.7% for CP and 16% for starch. In addition, Givens *et al* (2000b) also list several more references reporting large inter-laboratory variation for several analytes relevant in feed evaluation: OM digestibility, *in situ* rumen degradation, gas production kinetics, *in vitro* digestibility, ME and GE

In vitro and *in vivo* digestibility measurements have been shown to suffer from similar inter-laboratory variation. Madsen and Hvelplund (1994) sent 5 feeds to 23 laboratories in 17 countries and found "differences... in protein degradabilities between laboratories too large to be acceptable" when using nylon bags in the rumen of cows or sheep. The within laboratory variation, however, of protein degradation was acceptable. They concluded that sample preparation and processing and the bags themselves varied considerably between laboratories and made detailed recommendations for the nylon bag procedure.

Authors also recommend that a 'standard feed is made available for all laboratories for routine checking of analytical procedure'. A small ringtest of the Tilley-Terry method between 3 laboratories suggested that it was a robust method for OM digestibility of roughages (Madsen *et al*, 1997). However, problems were encountered with concentrates and highlights that careful standardization is required. Not surprisingly, feed manufacturers have moved to an enzyme method and such results are expected to be more reproducible (Madsen *et al*, 1997).

There are large differences between the different *in vitro* digestibility techniques (rumen fluid versus cellulase based techniques) which are used to predict *in vivo*digestibility (De Boever *et al*, 1994; Aufrère and Michalet-Doreau, 1988). Clearly, such differences in methods need to be resolved. These data collectively support the introductory statement that feed analysis can be difficult. Strict adherence to method details is important, especially if the method itself defines the component, *e.g.* crude fat, NDF, lignin, as empirical fractions are chemically not well defined (Bailey and Henderson, 1990). Zeeman and Bonn (1995) have also suggested that there should be an international definition of starch.

However, large variation may highlight problems in analytical methods and thus encourage closer investigation. Problems with too high NDF values, for example, are likely to be stem from incomplete solubilization of starch. Recently, Thiex *et al* (1996) investigated the observed large variation in reported vitamin A results, which was noted by the American Association of Feed Control Officials (AAFCO) who operate a feed check sample programme. Several recommendations were made to reduce errors of Vitamin A in animal feed and pet food analysis.

THE BENEFITS OF QUALITY ASSURANCE PROGRAMMES

A review of the literature for methods of feed grain analysis concluded that it was not possible to assess if variation in reported values was due to genotypic, environmental factors or inter-laboratory differences. Petterson *et al* (1999), therefore, recommended the use of quality assurance schemes, inter-laboratory evaluation programmes and reference materials.

The Global Environmental Monitoring Scheme of the WHO tested the performance of EU laboratories that contribute data on food contamination (Weigert *et al*, 1997). This involved 5 proficiency tests, 136 laboratories in 21 countries using their own preferred methods for the analysis of trace metals (Pb, Cd, Hg in milk powder), pesticides (organochlorine, organophosphorus, pyrethroid in spinach powder), nitrate in spinach powder, aflatoxins in nut-based animal feeds. Only 60% reported accurate for trace metals, 41% for pesticides, 43% for nitrate, 88% for aflatoxins and 53% for patulin (average = 68%).

Key *et al* (1997) summarised the results of the UK Food Analysis Performance Assessment Scheme (FAPAS) from 1990-1996. For pig feeds (moisture, ash, oil, protein, fibre, Cu), only 76% of laboratories achieved satisfactory results and for nutritional analysis 80% were satisfactory. The FAPAS study also reveals that some analyses are more difficult than others. 91% of aflatoxin data were satisfactory, 81% of veterinary drug residues, 86% of DDE, but only 71% of DDT and 65% of Ca data. Surprisingly, Ca analysis has shown little improvement over the years. Inter-laboratory comparisons and proficiency testing can highlight inappropriate methods, *e.g.* analysis of veterinary drug residues using certain immunological tests. However, once laboratories were participating in proficiency tests on a regular basis (*e.g.*

FAPAS) the average percentage for accurate results increased. Horwitz (1993) observed "most experimentation dealing with analytical methodology in biological sciences has been conducted within a single laboratory. Method validation by other laboratories was considered not only unnecessary but also detrimental because, in the words of one commentator, 'the results are too variable'. Within the last two decades, however... it has become increasingly apparent that a collaborative inter-laboratory study is the only way to estimate the variability characteristics of methods...." and to meet the increasing demand by regulatory programmes for high quality data.

How to Achieve Valid Data

The UK Department for Trade and Industry launched an initiative in 1994 on Valid Analytical Measurement (VAM) incorporating four main principles.

Measurements Should be Made Using Properly Validated Methods

Properly validated methods will provide information on the performance of an analytical technique, such as accuracy and precision, ruggedness, operating range, selectivity and limits of detection. It is essential when reporting a measured value to also give its uncertainty. Otherwise, it is not possible for users of the data to know what confidence to place in the data. It needs to be recognised that problems of communication can arise from the word 'uncertainty', as a layperson may misinterpret the statistical term 'error' into 'inaccurate data'! (Williams, 1996).

Guidelines are now available for the statistical evaluation of analytical tests and laboratory performance (Bailey and Henderson, 1990): the inter-laboratory precisions are generally a function of concentration. Horwitz' group found that the within-laboratory variation was approximately one-half to two-thirds of the between-laboratory variation and can be used as a 'bench mark for judging previously unevaluated methods' (Bailey and Henderson, 1990).

CRMs are used for:

- Calibration and verification of measurements under routine analysis conditions
- Internal quality control and quality assurance schemes
- Verification of the correct application of standardised methods
- Development and validation of new methods of measurement.

Unfortunately, only few CRMs are available for validation of analytical methods for feed analysis, especially for proximate analysis (Crosby, 1995). However, CRMs exist for minerals in two animal feeds: hay powder (Ca, K, Mg, P, S, Zn, I, N, Kjeldahl-N) and rye grass (As, B, Cd, Cu, Hg, Mn, Mo, Ni, Pb, Sb, Se, Zn). Information on the in-house production of reference materials can be found in Walker and Brookman (1998). It is good practice to include external CRMs or in-house reference materials into all analytical procedures.

PTS independently assess the performance of analytical laboratories. The true concentration of an analyte can be determined by addition of a known amount of analyte to a base material or better still through the use of a consensus value produced by a group of analysts. Guidelines exist for organising a collaborative study to evaluate analytical methods (AOAC, 1988). Appendix 2 lists several proficiency testing schemes.

It should be recognised that only certain accreditation schemes are appropriate for laboratories performing chemical analysis, *e.g.* NAMAS M10, ISO 90025, ISO/IEC 17025, EN 4501. General guidelines have been written on how to prepare for accreditation (EURACHEM/WELAC, 1998). Gangaiya and Morrison (1992) listed general problems of setting up quality assurance in some developing country laboratories. Appendix 2 lists several accreditation organizations.

SAMPLE PREPARATION

Good analytical data require that samples be representative of the whole and that their integrity has been ensured during transport to the laboratory and during their preparation (drying and grinding). Relevant guidelines for feeds can be found in Feeding Stuffs (1988), AOAC (1995), Crosby (1995) and Wrigley (1999). Drying may adversely affect the analysis of sugars, vitamins, certain trace elements (F, Se, B), and ammonia and VFAs in silages (MAFF, 1986). Caution needs to be taken when grinding lupin and chickpea seeds (Petterson *et al*, 1999).

Moisture

Several methods exist to determine moisture content of feeds: oven drying at 105°C for 16h, 125°C for 4h or 135°C for 3h (AOAC, 1995). Molasses should be dried at 70°C because of high levels of volatile compounds (Petterson *et al*, 1999). Oven-drying is problematic with silages and high fat feeds; vacuum-oven drying at 95-100°C (AOAC, 1995) or at <70°C, Karl Fischer or toluene distillation are alternative techniques (Crosby, 1995; Cherney, 2000). Baker *et al* (1994) found significant variation in moisture contents between laboratories, which were tracked down to variable temperature gradients in ovens (Givens, pers. communication).

Ash

Crude ash is determined either by ashing at 600°C for 2 h or between 500-550°C for 12-16 h (Midkiff, 1984; Petterson *et al*, 1999). For difficult samples, Crosby (1995) lists some special techniques. Flameless atomic absorption spectrometry on solid feed samples for trace levels of Cu represents an interesting development (Anzano *et al*, 1994). However, sample size at present is restricted to 2-4 mg. This technique, therefore, requires further development for general applicability in feed analysis.

Crude Protein

Crude protein data are standard for evaluating the protein value of forages (Cherney, 2000). The historic developments and critical points in Kjeldahl nitrogen measurements (choice of catalyst, temperature and digestion times) have been summarised by Lakin (1978) and Midkiff (1984). The Dumas technique represents an alternative method for total nitrogen. Total nitrogen is determined after combustion of the sample and several commercial instruments are now available for this. The measurements are rapid taking only 2-5 minutes per sample. The small sample size (20 to 500 mg) is the key problem with the Dumas method. Samples should be finely ground and preferably dry to avoid too frequent changes of costly reagents. Therefore, the best method for some materials, *e.g.* silages, is still the Kjeldahl procedure as drying would remove ammonia-N.

The need for corrosive and toxic reagents is the main disadvantages of the Kjeldahl method. Total Dumas nitrogen values can be slightly higher than the Kjeldahl values as Dumas N also includes nitrate and organic compounds that are highly resistant to acid digestion (Lakin, 1978; Petterson *et al*, 1999). As a result, exceptional differences can occur with some biological matrices, *e.g.* fruits, vegetables and fish, with Kjeldahl-N:Dumas-N ratios as low as 0.15 (Simonne *et al*, 1997).

A factor of 6.25 is used to convert total nitrogen in animal feeds into crude protein. However, the amino acid composition varies between foods and therefore different factors have been suggested to convert total N to crude protein (Petterson *et al*, 1999). It can range from 5.14 for grains and oilseeds to 6.38 for dairy products. As there is some doubt about the universal validity of such factors, authors should report crude protein values together with the factors used in their calculation (Lakin, 1978).

It should be noted that none of the dye-binding methods were universally acceptable between laboratories for protein determinations (Petterson *et al*, 1999). Recently, Strong and Duarte (1992) described a simple and rapid biuret method for protein determination in wheat, rice and soyabeans. This has been applied to a wide range of different grains and only requires a blender, reagents, colorimeter at 550nm and minimum operator training. The method requires initial calibration against protein values obtained by other techniques. This procedure may be of interest to grain and feed merchants as it can be completed in 5 minutes.

Fibre Analysis

It is important to recognise that all fibre determination employ 'empirical' methods, *i.e.* the method determines the final result and any deviation from the analytical protocol will produce a different result (Midkiff, 1984; Crosby, 1995). Midkiff (1984) summarised the history of crude fibre analysis. Cherney (2000) contrasted the Weende proximate analysis system, which originated in 19th century and has hardly changed since then, with the Van Soest system developed in the 1960s. He cautions against the use of CF, NFE, EE in feed evaluation systems, as they do not sufficiently separate digestible from non-digestible fractions. The Van Soest system is now widely used for forage evaluation as it provides useful measurements for nutritionally important parameters, such as structural carbohydrates (Goering and Van Soest, 1970; AOAC method 973.18).

Several groups have since modified the original manual method (Cherney *et al*, 1989; Van Soest *et al*, 1991) and Chai and Uden (1998) have described a simple oven-based procedure. The micro-NDF method by Pell and Schofield (1993) is another modification, which allows NDF to be determined on small sample sizes (10 to 50 mg). Moore and Hatfield (1994) comprehensively reviewed the composition and analytical methods for structural and non-structural carbohydrates citing research involving both ruminants and monogastric animals. Mertens (1997) concluded, "the only fibre method that can be used on all types of feeds is the method recommended by the National Forage Testing Association" (Undersander *et al*, 1993, Hintz and Mertens, 1996; see also Appendix 2 for web site).

Starch removal from NDF can be difficult and requires pre-treatment with α-amylase either overnight (McQueen and Nicholson, 1979) or by using heat-stable α-amylase (Sigma product A3306) during the last 30 minutes of NDF extraction and during filtration (Cherney *et al*, 1989). Difficult samples may also require pretreatment with 8M urea (Van Soest *et al*, 1991). Filtration problems may be due to incomplete starch removal. However, errors in fibre analysis may also stem from aged crucibles, as crucibles can be damaged at too high temperatures or by rapid rates of heating or cooling in muffle ovens (Crosby, 1995).

Fibretec (Tecator, Hoenganaes, Sweden) and the recent FibreAnalyzer (ANKOM Technology Co., Fairport, NY, USA) are instruments for NDF and ADF extractions. Studies indicated that ADF by Fibretec and FibreAnalyzer are comparable for most feeds. However, in our experience NDF in starch containing feeds is best analysed by the FibreAnalyzer as it removes starch more efficiently (unpublished data).

Starch

Starch consists essentially of two components, amylose and amylopectin. Amylose polymers contain up to 2000 glucose units connected through linear 1→ 4 linkages. Amylopectin is a highly branched polymer containing 2,000 to 220,000 glucose units with 1→ 4 and 1→ 6 linkages. Therefore proper extraction techniques are crucial for successful starch analysis (Petterson *et al*, 1999). Starch is determined by pretreating with 80% ethanol in boiling water to remove low molecular weight sugars, followed by gelatinization and solubilization, before extracting and hydrolyzing the starch (Faichney and

White, 1983; Åman and Graham, 1990; Hall, 1997). Incomplete dissolution and incomplete accessibility to enzymes tend to be the main problems in starch analysis. McCleary *et al* (1994; 1997) reported that a ringtest involving 29 laboratories to evaluate the Megazyme enzyme kit for cereal products and some animal feeds produced good agreement for total starch content. This enzyme kit was accepted as an AOAC method (996.11) (Petterson *et al*, 1999).

The proportion of amylose and amylopectin depends on the source of the starch. Starch digestibility is not necessarily related to total starch content as the amylose:amylopectin ratio and processing affect the extent of starch digestion (Allen *et al*, 1997); Reynolds *et al*, 1997). The Rapid Visco Analyser measures pasting properties of starch in grain and Wrigley (1999) speculated that this may be related to starch digestibility, but this awaits further research.

Crude Fat

Fat in plant-derived feeds consists mainly of mono-, di- and triacylglycerides, free fatty acids and phospholipids. Feeds also often contain fats from animal and other waste products. Depending on the rendering processes used, heating or storing can lead to unsaponifiable matter, oxidised and polymerised fatty acids which will contribute to crude fat values, but not be of nutritional value (Edmunds, 1990). Midkiff (1984) described the history of crude fat analysis and the sample types that caused problems.

Several official methods exist to determine crude fat in animal feeds (AOAC 945.16 and 920.38; MAFF1986). Crude fat methods are empirical methods and procedural details must be closely adhered to. They are based on solvent extractions with or without hydrolysis. Crude fat is extracted by the EU procedure A with petroleum ether (40-60°C) and the dried residue weighed (Feeding Stuffs 1988). Acid hydrolysis is used as a pretreatment in EU procedure B. However, it is recommended that animal feeds are first extracted with petroleum ether (40-60°C) (Procedure A) and then subjected to acid hydrolysis, before re-extraction with petroleum ether (40-60°C) (Procedure B). This minimizes losses during filtration of the acid digest, which can occur if protected fats, *i.e.* Ca and Mg salts of fatty acids are present (Edmunds, 1990; Crosby, 1995).

Neither procedure A nor B will give satisfactory results if feeds contain milk products. The Rose-Gottlieb method is required for such feeds, as the alkaline pretreatment frees occluded lipids from protein capsules (ISO 1211:1984). Similarly, canned dog foods require extraction with a series of solvents after hydrolysis (Budde, 1952; Midkiff, 1984; AOAC 954.02). Problems may arise if diethyl ether is used instead of petroleum ether as suggested in some methods. Slightly higher values may result if water has not been removed completely from either the sample or the solvent, as some compounds, such as urea and sugars, are slightly soluble in diethyl ether in the presence of small amounts of water (Midkiff, 1984; AOAC 920.39). Great care is required when performing solvent extractions. Solvent are usually recycled between a lower electric heating source and an upper water-cooling system in a Soxhlet apparatus. Accidents are not uncommon with such a set-up. For this reason, a different type of extractor has been developed recently (Brown and Mueller-Harvey, 1999). The Soxflo instrument requires neither heat nor cooling water and is based on a dry-column procedure. The sample is packed into a column and the extracting solvent drips slowly through it. Crude fat values determined by the Soxflo and Soxhlet procedures were found to be in close agreement. If sample drying and packing procedures are evaluated against the Soxhlet procedure, good and reliable crude fat data can be obtained in 60 to 90 minutes.

Secondary Plant Products: Tannins

Although a whole range of different plant secondary products exists, only tannins will be discussed here. The main reason for this is the persistent confusion that surrounds tannins in animal nutrition. Analytical techniques have often been misapplied and data misinterpreted. Reference will however be made to rapid analyses of mycotoxins and other secondary plant products which can contaminate or affected the nutritive value of animal feeds.

Tannins comprise a diverse group of phenolic compounds, varying in molecular size from 500 to possibly 28,000 Daltons (Jones *et al*, 1976; Mueller-Harvey and McAllan, 1992; Schofield *et al*, 2001). Some are easily extracted by aqueous solvents, others are not and can be measured as fibre or protein bound tannins (Jackson *et al*, 1996; Reed, 1986). The naming and thereby classification of condensed (CT) or hydrolysable tannins (HT) is somewhat misleading as some of the 'condensed' tannins are relatively easily degraded oxidatively, whereas some of the 'hydrolysable' tannins will resist all attempts at hydrolysis (Mueller-Harvey, 1999; 2001).

Negative and positive animal responses have been attributed to tannins, *i.e.* ranging from animal death to increased growth rates (Butter *et al*, 1999; Mueller-Harvey, 1999). Despite many animal studies involving tanniniferous feeds, little attempt has been made to elucidate the relationship between animal production and tannin structures. Progress is unlikely to be achieved by the continued use of colorimetric tannin assays alone (Petterson *et al*, 1999). Lowry *et al* (1996) put it succinctly: 'the simplicity of absorbance measurements masks the problems of extracting meaningful data'.

A general reagent produces varying colour yields from different phenolic compounds, which includes tannins (Folin and Ciocalteau, 1927). However, phenolics and tannins tend to occur in mixtures and quantitation is not possible unless isolated standards are used, which are representative of the exact composition of the material being examined. Procedures for colorimetric methods for CT and HT have been described by Graham (1992), Waterman and Mole (1994) and Hagerman (web site). Porter *et al* (1986) recommended a modification of the butanol-HCl assay. This assay is excellent for detecting the presence of CT. However, for quantitation purposes it should only used if isolated tannins are used as calibration standards as tannin structures have a marked effect on colour yield (Giner-Chavez *et al*, 1997). Even within the same species, there can be sufficient structural variation to warrant isolated tannin standards from different accessions in quantitative work (Stewart *et al*, 1999). More recently, a CT assay based on acidified 4-dimethylamino-cinnamaldehyde (DMACA-HCl) was described (Li *et al*, 1996). This assay proved more sensitive than the vanillin-HCl method, which can suffer from interference by water (Terrill *et al.*, 1990). The DMACA assay was also used successfully on TLC plates and as a histochemical assay. This method has not yet been examined thoroughly against interferences from monomeric flavanols.

The analysis of HT has been reviewed by Mueller-Harvey (2001). Again, assays are best used to detect HT, as accurate quantitation would require isolated standards (Hagerman *et al*, 1997). Furthermore, there are just a few known species that produce only gallotannins, from which gallic acid can be hydrolysed before detection (Willis and Allen, 1998). More usually, gallo- and ellagitannins occur together as mixtures. Ellagitannins are much more difficult to measure (Wilson and Hagerman, 1990) although the free ellagic acid can be measured by HPLC (Mueller-Harvey *et al*, 1987). However, as stated above, many other so-called HT do not release gallic or ellagic acid.

It is recommended that additional assays be used in tannin analysis to overcome some of the problems of the colorimetric tests, *e.g.* the Yb-precipitation method (Reed *et al* 1985; Giner-Chavez *et al*, 1997; Krueger *et al*, 2000a) and/or one of the protein or polymer binding assays (Makkar *et al*, 1987; Dawra *et al* 1988; Makkar *et al* 1993) when screening novel fodder plants. If facilities are available to measure radioactivity, tannins can also be measured by binding to ^{125}I labelled BSA (Hagerman and Butler, 1980) or ^{14}C-PEG (Silanikove *et al*, 1996). Tannins can be estimated directly in feed samples, without prior extraction in the latter method.

The Yb-method precipitates total phenolics (*i.e.* tannins, flavonoids and other phenolics) which correlate positively with butanol-HCl CT (Reed *et al*, 2000). The advantage is that phenolics are determined by gravimetry without the need for standards. Interestingly, the amount of PEG bound per gram of sample correlated highly significantly with*in vitro* N digestibility and appeared to be a more meaningful measurement than the colorimetric asssays (Jones and Palmer, 2000). Correlations with CT measurements by vanillin-HCl were better than by butanol-HCl. In view of the problems of colorimetric assay, it seems surprising that thin layer chromatography (TLC) has not been used more widely. TLC reveals the presence of CT vs HT, low vs high molecular weights of CT, the subclasses of CT or HT (procyanidins or prodelphinidins, gallo- or ellagitannins), plus semiquantitative information based on colour intensity (Mueller-Harvey *et al*, 1987). Tannins from different species have also been compared by reverse phase HPLC (Mueller-Harvey at al 1987), size exclusion chromatography (Yanagida *et al*, 1999; Hedqvist *et al*, 2000) and normal phase chromatography (Tanaka *et al* 1984; Okuda *et al*, 1989; Hagerman *et al*, 1992). Okuda *et al* (1989) reviewed the mass spectrometric analysis of tannins. More recently, ESI-MS (Guyot *et al*, 1997; Marais *et al*, 2000) was used to determine tannin molecular weights. New developments in MALDI-TOF mass spectrometry (Hedqvist *et al*, 2000; Krueger *et al*, 2000a and b) succeeded in determining the molecular weights and composition of complex tannin mixtures.

Immunoassay Techniques

Although mycotoxins pose severe hazards to humans and animals, there were still a large number of countries in 1987 that had no specific regulations on mycotoxins (Van Egmond, 1993). Mycotoxins in feeds, along with many other organic compounds, can be analysed by either instrumental (HPCL, GC, CE) or biochemical (immunoassays) techniques. Skerritt and Appels (1995) described the basic principles of enzyme-linked immunosorbent assays (ELISAs) and illustrated the different forms of techniques (direct or indirect assays, sandwich or competition assays).

For example, in the direct competitive ELISA format, enzyme-linked antibodies (E-As) coat the surface of a well plate or test tube, which are supplied in a test kit. The operator extracts the feed sample and places the extract onto the surface. Compounds of interest, *i.e.* antigens, bind to the E-As thereby releasing them from the surface. E-As are then washed away and an enzyme substrate is added to the remaining, bound E-As. This procedure results in an inverse relationship between antigen concentration and colour production: the higher the antigen concentration, the more E-As have been washed away and the smaller the colour yield of the reaction.

Several commercial kits are available for aflatoxins, zearalenone and other mycotoxins, alkaloids, glucosinolates, insecticides, herbicides, fungicides, various environmental pollutants, vitamins in foods and animal feeds. Morgan (1995) reviewed these ELISA techniques and listed suppliers' addresses. Schneider *et al* (1991) developed an interesting dipstick technique for the simultaneous detection of several mycotoxins.

ELISAs can be used by regulatory authorities, quality control laboratories and in research laboratories. They require relatively little user training and can be used in small laboratories or under field conditions. Low cost and high speed make ELISAs ideal for on-site monitoring of stored grain and for assessing rapidly if the maximum residue levels of traded animal feeds have been exceeded. They can also be used for checking suspected spillages. High sample throughput facilitates elimination of large numbers of negative, uncontaminated samples. It is, however, recommended that samples giving positive results for a contaminant should be re-analysed by conventional instrumental techniques to ensure absence of matrix interference and accuracy of the data (Petterson *et al*, 1999). Some commercial test kits compare well with standard AOAC methods for aflatoxins in animal feeds (Trucksess *et al*, 1989 and 1990; Cochrane, 1991) and can be used for quantitative or semiquantitative measurements in as little as 3 minutes. Several of these have now obtained AOAC approval (Trucksess *et al*, 1989).

ELISAs can be tailored to be selective for individual compounds or compound classes, parent compounds or metabolites and even for isomers. Highly sensitive ELISAs for M_1 in milk samples have been reported (Kawamura *et al*, 1994). Of interest in the current European BSE crisis is a report by von Holst *et al* (2000), who evaluated the applicability of a commercial ELISA method to detect proper heat treatment of pork and beef meals. A word of caution. ELISA procedures - like any analytical method - need to be carefully evaluated as some solvents used for analyte extraction may adversely affect antibody performance (Morgan, 1995). In an inter-laboratory study, ELISA tests gave good results for ZON, but accuracy with the DON kits was poor (Schuhmacher *et al*, 1997).

Matthews *et al* (1996) investigated commercial kits for three organophosphorus pesticides (OP) for laboratory and field use. The authors found good correlation with GC methods (r-values between 88 and 98%). However, as the protocols supplied with the kits were somewhat confusing, the authors provided improved protocols for use by grainstore keepers, millers and malsters. It was also found that the operating ranges of the kits were not as wide as claimed by the manufacturer. In addition, some cross-reactivity of the antibodies was observed with structurally related compounds. Nonetheless, the study showed that ELISA tests could be used reliably to measure OP residues on stored grain.

The most common problems encountered in ELISAs are incomplete washing steps, pipetting problems, insufficient temperature control of reagents and plates, degradation of conjugate (may inhibit good binding to antibody) or loss of enzyme activity (Gee *et al*, 1995). Problems with sample matrices are also common and can be detected by comparison of standards in an uncontaminated or 'blank' matrix.

Recovery studies of standard additions may show up matrix problems. Alternatively, a dilution curve with sample matrix solution can be compared against a standard calibration curve. If the slopes are not parallel, a matrix effect is likely. Finally, when purchasing an assay kit, the following criteria should be considered: price per assay, sensitivity, cross-reactivity, suitability for the chosen matrix, availability of published validation by independent workers and technical support (Gee *et al*, 1995).

Near Infrared Reflectance Spectroscopy

Deaville and Flinn (2000) wrote a brief and clear introduction to the basic principles of NIRS. An accessible and more detailed review of applications and a description of the mathematical treatments of NIR spectra has been given by Shenk and Westerhaus (1995). Givens and Deaville (1999) have also reviewed NIRS in relation to feed analysis and animal nutrition.

The advantages of NIRS (Givens and Deaville, 1999) over traditional techniques are:

- Rapid, as minimal or no sample preparation necessary
- On-the-spot analysis of the whole sample, *i.e.* a non-destructive technique, allows simultaneous measurement of several parameters
- High precision
- High throughput makes NIR a cheap technique on a sample basis
- Environmentally friendly: no reagents, no chemical waste

The limitations of NIRS are:

- Suitable for major feed components, not for minor components
- Great care needed in developing calibrations as these are matrix specific
- Complexity in choice of data treatment is confusing for the novice
- Calibration procedures are time consuming and only worthwhile for subsequent analysis of large sample numbers
- High instrument costs

NIRS was developed in the 1950s and 1960s for quantitative analysis and applied to feed analysis in the 1970s (Norris *et al*, 1976). By the late 1970s, NIRS was routinely used for protein measurements in grain. The NIRS region covers wavelengths between 730 and 2500 nm and is the infrared region that is particularly suited to quantitative analysis (Givens and Deaville, 1999). The main absorption bands of water are at 1940 and 1450 nm, of aliphatic C-H bonds at 2310, 1725, 1400 and 1210 nm, of O-H bonds around 2100 and 1600 nm and N-H bonds at 2180 and 2055 nm. Spectra are depicted in the form of the reciprocal log of reflectance (log1/R) and provide little direct information. Various components in the feed matrix produce a series of overlapping bands, which results in a smooth, rolling line. However, first- and second-order derivatives of the log1/R spectra can be used to resolve these overlapping bands.

NIRS data are generally subjected to a mathematical pretreatment to reduce interferences from light scatter (Barnes *et al*, 1989). Then one of several different multivariate calibration methods is used to relate the spectral data from a sufficiently large and representative sample set to the primary, 'wet chemistry', data (Blanco *et al*, 1997). Finally, calibrations are subjected to validation procedures with an independent set of samples. A simple monitoring procedure has been developed to minimise NIRS analysis errors (Shenk *et al*, 1989). Great care should be taken in developing NIR calibrations (references in Deaville and Flinn, 2000). Calibrations should be based on at least 50 samples, but often many more are required (>150). Givens and Deaville (1999) pointed out that 'NIRS is largely a secondary technique requiring calibration using samples of known composition determined by using standard methods (primary techniques)'. A problem can arise, when the primary methods do not define well the chemical constituent, *e.g.* drying at 100°C to determine moisture does not necessarily define water content or 6.25 x total N does not necessarily describe protein content adequately (Shenk and Westerhaus, 1995).

It has been stressed that consistent sample preparation is required, as variations in particle size, residual moisture content and packing density can adversely affect NIR spectra. However, new developments of NIR software, such as the noise repeatability file, have succeeded in reducing the sensitivity to residual sample moisture (Baker *et al*, 1994; Shenk and Westerhaus, 1995). Alternatively, by using a coarse transport cell, a larger surface area of fresh grass silage can be screened, thus eliminating the need for dry silages (Park *et al*, 1999a, b). NIRS has been accepted as an official AOAC method for crude protein and ADF (AOAC 989.03) and for moisture (AOAC 991.01; Barton and Windham, 1998). It has also been used for determining starch and non-starch polysaccharides, fat and oil, metabolisable energy, insect or weed seed contamination in feed grains (Wrigley, 1999) and for the analysis of dried forages (Murray, 1993). It can be used to identify feeds and perform authenticity checks (De Boever *et al*, 1993). In addition, heat damaged protein, fungal contamination and adulteration can be detected with modern pattern recognition software (Givens and Deaville, 1999).

It is also possible to transfer calibrations developed on an expensive scanning instrument to cheaper filter instruments in a network (Puigdomènech *et al*, 1997). This ISI cloning software also allowed a successful transfer of NIR calibrations developed for fresh grass silages from a Foss to a Bran & Luebbe instrument (Park *et al* 1999b).

Researchers are now aiming to predict directly the functional properties of feeds to animals, *i.e.* nutrient supply and production responses such as live weight gain, milk fat and protein or meat composition, rather than measuring feed components (Wrigley, 1999). Good predictions have been achieved for organic matter digestibility *in vivo*(Barber *et al*, 1990), metabolisable energy (ME) content (Givens *et al*, 1992) and voluntary feed intake. It has also been possible to predict nutritionally relevant products, such as lactic acid, VFAs and cumulative gas volumes with calibrations based on 800 fresh silages (Deaville and Flinn, 2000). The ultimate aim is to formulate diets for optimum animal productivity, cost effectiveness and the least environmental effects (Givens and Deaville, 1999).

ANALYSIS OF FEED

In order to formulate, produce and market animal diets it is necessary to gain information about substance classes (i. e. that contribute to the energy content) and special compounds (i. e. that influence digestibility). Different chemical, physical and biological methods have been developed to obtain the information needed.

Sampling

The most important precondition when analyzing feedstuffs and diets is the sampling procedure. Thereby the person taking the samples has to keep the following principles in mind:

- The sample taken must be representative for the entire lot.
- Samples are to be taken randomly from several different points of the lot. Subsequently the samples are then mixed to a single blend to produce a collective sample, which again is divided into several representative laboratory samples for analysis.
- Sampling equipment should not be able to influence the sample taken, i. e. via contamination or sedimentation.
- Feed samples need to be stored in a manner that ingredients will not be altered (temperature, oxygen, sun light etc.) before analysis.
- A sampling report should be prepared in order to assign the sample correctly after analysis.

Proximate Analysis (Crude Nutrient Analysis)

The proximate or Weende analysis of feed is a quantitative method to determine different macronutrients in feed. Basically it is the partition of feed compounds into six categories by means of common chemical properties. The categories are moisture (crude water), crude ash (CA), crude protein (CP), ether extracts (fats or lipids; EE), crude fibre (CF) and nitrogenfree extractives (NFE).

The feed sample is initially dried at 103 °C for 4 hours. The weight loss of the sample is determined and the crude water fraction is calculated. Ashing the sample at 550 °C for 4 hours removes the carbon from the sample, *viz*. all organic compounds are removed.

Again calculating the weight loss of the feed sample from the dry matter to crude ash (CA) content mathematically determines the organic matter fraction. The nitrogen content of the food is the basis for calculating the crude protein (CP) content of the feed. The method established by Kjeldahl converts the nitrogen present in the sample to ammonia which is determined by titration.

Assuming that the average nitrogen content of proteins is 16 per cent multiplying the nitrogen content in per cent obtained via Kjeldahl analysis with 6.25 gives an approximate protein content of the sample. Alternatively the Dumas method can be applied to measure CP. Fats and lipids are extracted continuously with ether, after evaporation of the solvent the residue remaining is the ether extract (EE) fraction. The carbohydrates in a feed sample are retrieved in two fractions (CF, NFE) of the proximate analysis. The fraction, which is not soluble in a defined concentration of alkalis and acids, is defined as crude fibre (CF). This fraction contains cellulose, hemicellulose and lignin. Sugars, starch, pectins and hemicellulose etc. are defined as nitrogen-fee extractives (NFE). This fraction again is not determined chemically it is rather calculated by substracting CP, EE and CF from organic matter.

In recent years the over 100 year old proximate system has been advanced and improved. Especially the imprecision of CA, CF and NFE as well as CP had been criticized. Modern methods to determine the exact composition of the CA fraction via atomic absorption spectroscopy and the CP fraction via amino acid analyzers, near infrared spectroscopy (NIRS) etc. have been established. Improving the information gained from analysis of feedstuffs and diets also involves the determination of sugars and starch (polarimetric methods) contained in the NFE fraction of the proximate analysis. Van Soest developed a procedure to detect the different components of the cell wall. This helps specify the CF and NFE fraction.

Thereby the complete amount of cell wall components is obtained by digesting (boiling) the feed sample in a so called neutral detergent solution and results in the neutral detergent fibre fraction (NDF). The residue after digestion in a solution with sulfuric acid is called the acid detergent fibre (ADF) and contains mainly cellulose and lignin. Finally the remaining sample is treated with a sulfuric acid with an even higher concentration resulting in a decomposition of cellulose leaving mainly lignin. This fraction is called acid detergent lignin (ADL). The combination of the proximate analysis with these modern methods allows for a detailed feedstuff analysis.

UNDERSTANDING FEED ANALYSIS

Feed costs represent the largest annual operating cost for most commercial cow-calf enterprises. In order to maintain an optimum balance between feed costs and production, feeds must be analyzed and these analyses used to formulate rations and (or) supplements. Feedstuffs vary widely in nutrient concentration due to location, harvest date (maturity), year, and other management practices. Tabular values may be used if necessary, but it is important to remember that they are average values and that significant variation exists. On a dry matter basis, energy can easily vary ±10%, crude protein ±15%, and minerals by a much greater margin.

Once a feed sample has been collected properly, it can be analyzed for nutrients. Most commercial laboratories offer standard feed tests for forages, grains, or total mixed rations. Analyzing cattle feeds for moisture, protein, and energy is recommended. Furthermore, you may wish to identify key minerals or minor nutrients of interest. Typically, results are reported on an as-is and dry matter basis. Nutrients should always be balanced on a dry-matter basis because nutrient requirements for beef cattle are reported on a dry-matter basis. After formulation on a dry-matter basis, values can be converted to an as-is basis (using the moisture content of the feed) to determine the actual amount of feed (as-is) that should be fed. Feedstuffs can be analyzed using traditional wet chemistry technique or near infrared reflectance spectroscopy (NIR). Samples can be analyzed more quickly, and usually cheaper, using NIR. However, NIR is only useful for feedstuffs and ingredients that have been well characterized using wet chemistry. Therefore, be sure to ask the laboratory if their database for your particular sample is extensive enough to ensure accurate results, particularly if you are analyzing less common feedstuffs.

MOISTURE

- *Dry Matter* (DM): Dry matter is the moisture-free content of the sample. Because moisture dilutes the concentration of nutrients but does not have a major influence on intake (aside from severe deprivation), it is important to always balance and evaluate rations on a dry-matter basis. Digestible Dry
- *Matter* (DDM): Calculated from acid detergent fibre (ADF; see below); the proportion of a forage that is digestible.

PROTEIN

- *Crude Protein* (CP): Crude protein measures the nitrogen content of a feedstuff, including both true protein and non-protein nitrogen. In ruminants, evaluation of the fraction that is degradable in the rumen, degradable intake protein (DIP), versus the rumen-undegradable fraction, undegradable intake protein (UIP), is also important. However, the rumen degradability of protein is not measured in most commercial labs. Therefore, it is recommended that rations be formulated using analyzed CP values and average values for DIP and UIP that can be found in the 1996 National Research Council Nutrient Requirements of Beef Cattle.
- *Degradable Intake Protein* (DIP): The fraction of the crude protein which is degradable in the rumen and provide nitrogen for rumen microorganisms to synthesize bacterial crude protein (BCP) which is protein supplied to the animal by rumen microbes. DIP also includes non-protein nitrogen found in feeds or ingredients.
- *Undegradable Intake Protein* (UIP): The rumen-undegradable portion of an animals crude protein intake. Commonly called "bypass protein" because it bypasses rumen breakdown and is mainly digested in the small intestine. Bypass protein is utilized directly by the animal because it is absorbed as small proteins and amino acids.
- *Metabolizable Protein* (MP): MP is protein that is available to the animal including microbial protein (BCP) synthesized by the rumen microorganisms and UIP.
- *Heat Damaged Protein or Insoluble Crude Protein* (ICP): Nitrogen that has become chemically linked to carbohydrates and thus does not

contribute to either DIP or UIP supply. This linkage is mainly due to overheating when hay is baled or stacked with greater than 20% moisture, or when silage is harvested at less than 65% moisture. Feedstuffs with high ICP are often discolored and have distinctly sweet odours in many cases. When the ratio of ICP:CP is 0.1 or greater, meaning more than 10% of the CP unavailable, the crude protein value is adjusted. Adjusted crude protein (ACP; see below) values should be used for ration formulation.

- *Adjusted Crude Protein* (ACP): Crude protein corrected for ICP. In most nutrient analysis reports, when ACP is greater than 10% of CP, the adjusted value is reported. This value should be used in formulating rations when ICP:CP is greater than 0.1.
- *Digestible Protein* (DP): Reported by some laboratories, do not use without the guidance of a nutritionist. Digestible protein values are not needed for most ration formulation because nutrient requirements and most formulation tools are already adjusted for protein digestibility. Furthermore, protein digestibility is influenced by external factors.

FIBRE

- *Crude Fibre* (CF): Crude fibre is a traditional measure of fibre content in feeds. Neutral detergent fibre (NDF) and acid detergent fibre (ADF) are more useful measures of feeding value, and should be used to evaluate forages and formulate rations.
- *Neutral Detergent Fibre* (NDF): Structural components of the plant, specifically cell wall. NDF is a predictor of voluntary intake because it provides bulk or fill. In general, low NDF values are desired because NDF increases as forages mature. Acid Detergent Fibre (ADF): The least digestible plant components, including cellulose and lignin. ADF values are inversely related to digestibility, so forages with low ADF concentrations are ususally higher in energy.

ENERGY

- *Total Digestible Nutrients* (TDN): The sum of the digestible fibre, protein, lipid, and carbohydrate components of a feedstuff or diet. TDN is directly related to digestible energy and is often calculated based on ADF. TDN is useful for beef cow rations that are primarily forage. When moderate to high concentrations of concentrate are fed, net energy (NE, see below) should be used to formulate diets and predict animal performance. TDN values tend to underpredict the feeding value of concentrate relative to forage.
- *Net Energy* (NE): Mainly referred to as net energy for maintenance (NEm), net energy for gain (NEg), and net energy for lactation (NEl). The net energy system separates the energy requirements into their

fractional components used for tissue maintenance, tissue gain, and lactation. Accurate use of the NE system relies on careful prediction of feed intake. In general, NEg overestimates the energy value of concentrates relative to roughages.

- *Ether Extract* (EE): The crude fat content of a feedstuff. Fat is an energy source with 2.25 times the energy density of carbohydrates.
- *Relative Feed Value* (RFV): A prediction of feeding value that combines estimated intake (NDF) and estimated digestibility (ADF) into a single index. RFV is used to evaluate legume hay. RFV is often used as a benchmark of quality when buying or selling alfalfa hay. RFV is not used for ration formulation.
- *Relative Forage Quality* (RFQ): Like RFV, RFQ combines predicted intake (NDF) and digestibility (ADF). However, RFQ differs from RFV because it is based on estimates of forage intake and digestibility determined by incubating the feedstuff with rumen microorganisms in a simulated digestion. Therefore, it is a more accurate predictor of forage value than RFV. Neither RFV nor RFQ are used in ration formulation.

10

Understanding Ruminant Nutrition of Animals

In order to develop feeding systems, it is necessary to relate information on the nutritional characteristics of feed resources to the requirements for nutrients, depending on the purpose and rate of productivity of the animals in question. In the industrialized countries, this information has been incorporated in tables of "feeding standards" which interpret chemical analyses of feed resources in terms of their capacity to supply the energy, amino acids, vitamins and minerals required for the particular productive purpose. These standards are steadily becoming more sophisticated with the aim of improving their effectiveness in predicting rates of performance of intensively-fed livestock and to derive least cost formulations.

FEED COST OF FEEDING SYSTEMS

Feed costs account for 45 to 60% of milk production costs for confinement-based feeding systems. As a result, many dairy producers have become engrossed in reducing costs to feed a cow per day rather than addressing feed efficiency. *The least expensive ration is not usually the most productively-efficient ration*. This statement may sound like a contradiction, but relates to the understanding of how the cow and her rumen interact from a nutrient requirement perspective.

Nature has instilled a marvelous symbiotic relationship between bacteria and cow that allows the cow to consume material, which would be indigestible to the cow alone, and produce high quality meat and milk products. Dairy producers need to take full advantage of this cow-rumen interrelationship in order to produce milk and meat most efficiently and to minimize cow nutrition-related health problems.

If you are not taking advantage of the rumen, then you might as well be feeding pigs! To understand how to best meet cow and rumen nutrient requirements, one needs to appreciate the dynamic interaction of feeds and their components in the rumen as it impacts availability of substrate to support microbial and cow production. In understanding such a system, a new complexity to feed analysis must be defined. The focus of this lecture is to acquaint you with the specific nutrient requirements of the cow and her rumen and how they are appropriately met in an effort to produce milk as efficiently as possible.

RUMINANT NUTRITION

The ruminant nutrition programme is characterized by highly productive individual research programmes and a concomitant commitment to the pursuit of collaborative research efforts. Scientists within the ruminant nutrition programme maintain the dual goals of conducting research which will advance the understanding of fundamental nutritional phenomena and provide insight into practical aspects of the nutritional management of ruminant livestock. Students entering the programme are provided with a strong foundation in ruminal and post-ruminal digestion, absorption and metabolism as well as training in the fundamental experimental procedures necessary for conducting ruminant nutrition research.

Supporting coursework is frequently pursued in the areas of biochemistry, grain science, microbiology, physiology and statistics.

Areas of research emphasis within the ruminant nutrition group include:

- Dairy cattle nutrition
- Cow-calf nutrition
- Feedlot nutrition
- Growth and development
- Rumen microbiology (via Dr. Nagaraja in Vet. School)
- Stocker nutrition

UNDERSTANDING THE RUMEN

Ruminant herbivores, including cattle, sheep, goats, deer and many others, are unique animals in their ability to derive nutrients from forages and low quality roughages. This ability has nothing to do with the animal's digestive enzymes, but totally dependent upon the symbiotic relationship between host animal and microbial populations residing in the pregastric fermentation system called the rumen. What separates ruminant herbivores from non-ruminant herbivores is their ability to chew their cud. The ability to regurgitate swallowed feed material for remastication provides the rumen microbes greater surface area to feed materials thus allowing greater extent of degradation. Ruminant animals are the most efficient fibre digesting herbivores.

If the rumen system is functioning properly, it can provide a large proportion of the needed nutrients to support productive activities of the host animal. The rumen microbial population converts consumed dietary substances into highly available microbial protein and volatile fatty acids (VFA) that can be used by the host animal for protein synthesis and energy needs, respectively.

In addition to the rumen microbial end products, a proportion of dietary protein, fat, and starch may escape microbial degradation and be directly available for digestion by the host animal. Dietary nutrients that escape rumen degradation are commonly termed bypass protein, fat, or starch. It is the combination of rumen degradation products and dietary bypass nutrients that support all body functions of the host animal. The feeding programme becomes most efficient when microbial products from rumen degradation can account for a greater proportion of host animal needs and minimize the need for additional dietary bypass nutrients.

To accomplish this, one needs to fully understand the interactions of dietary substrates with rumen environment and impact on fermentation end product production. Within the last decade much research has focused on developing models that can accurately predict rumen output based on dietary inputs. To this end, the Cornell Net Carbohydrate and Protein System (CNCPS) has provided great insight as to the inner workings of the rumen and thereby increased the potential for improved and more efficient dietary formulation.

TANNIN TOXICITY IN RUMINANT ANIMALS

Tannins are phenolic compounds that are commonly found in plants. Found in the leaf, bud, seed, root, and stem tissues, tannins are widely distributed in many different species of plants.

Tannins are separated into two classes: hydrolysable tannins and condensed tannins. Depending on their concentration and nature either class can have adverse or beneficial effects. Tannins can be beneficial, having been shown to increase milk production, wool growth, ovulation rate, and lambing percentage, as well as reducing bloat risk and reducing internal parasite burdens.

Tannins can be toxic to ruminants, in that they precipitate proteins, making them not available for digestion, and they inhibit the absorption of nutrients by reducing the populations of proteolytic rumen bacteria. Very high levels of tannin intake can produce toxicity that can even cause death. Animals normally consuming tannin-rich plants can develop defensive mechanisms against tannins, such as the strategic deployment of lipids and extracellular polysaccharides that have a high affinity to binding to tannins.

Rumen Development

The above described rumen fermentation system is a marvelous system, however, one needs to recognize that ruminant animals are not born with a functioning rumen. The rumen system needs to develop anatomically and physiologically before it is capable of accomplishing extensive fermentation of feeds. Both anatomic size and muscular motility as well as physiologic development are dependent upon the type of diet fed the young preruminant calf. Relative size changes shown in Table depict gradual increasing rumen and reduction in abomasal sizes. These changes are based on optimum feeding programmes.

Table. Relative Changes in Anatomic Size of the Four Compartments of the Rumen System associated with Animal Age.

Compartment	Newborn	2-3 Months	Adult
Reticulum	5 %	5 %	5 %
Rumen	25 %	65 %	80 %
Omasum	10 %	10 %	7-8 %
Abomasum	60 %	20 %	7-8 %

In addition to anatomic changes, fermentation ability of the rumen needs to be developed. First, microbial organisms must be inoculated into the rumen and then the absorptive surface must anatomically grow and develop metabolic activity. Rumen microbial inoculation occurs very quickly with the calf's exposure to feeds, environment and other animals. This is usually never a problem under normal situations.

The second component of rumen physiologic development is papillae growth and metabolic activity. Historically, it has been believed that consumption of forage by the calf would initiate rumen papillae development. Research work from the 1960's as well as most recently has definitively shown that the end products of starch fermentation, namely butyrate and propionate, are the mediators of rumen papillae development. A number of studies have shown how rumen papillae development can be suppressed by continued milk feeding without sufficient calf starter intake.

What this means is the feeding of calf starter is critical to the rapid development of proper rumen function and starter consumption by the calf should be encouraged. Contrary to popular belief, feeding of forages will not promote rumen development, but will stimulate rumen muscular activity. Fermentation of forages generates predominately acetate, which has little stimulatory activity on papillae development. For calves to make a smooth transition from the milk feeding phase through weaning, rumen development should be well initiated, which can be accomplished when calves are eating 700 to 900 g of calf starter per day for at least 3 consecutive days prior to weaning.

Applied Rumen Anatomy

The rumen is actually only one chamber of a complex, pregastric fermentation system. This is in contrast to the postgastric fermentation system found in horses and many other non-ruminant herbivores. The reticulum is a smaller fermentation compartment, anterior and intimately associated with the ruminal compartment. The reticulum is primarily responsible for assisting in rumination contractions and distributing feed within the reticulo-rumen. You may be more familiar with this compartment from its association with "hardware disease". Heavy objects that are swallowed by the cow will drop into the reticulum.

This compartment is located next to the heart, just the other side of the diaphragm. Sharp objects may protrude through the reticulum wall and puncture the heart or liver, thus causing "hardware disease". The rumen is the primary fermentation vat, being between 80 and 100 liters in volume in a mature cow. Muscular contractions aid in the constant mixing of feed materials with bacteria laden fluids to promote fermentation and in the regurgitation of feed materials, which results in particle size reduction from chewing and stimulates copious production of saliva. Salivary bicarbonate ion is primarily responsible for maintaining only a slightly acid pH in the rumen, given the tremendous amount of acids being produced during fermentation.

Table. Characteristics of the rumen environment.

pH	6.7 - 7.2 optimum
Temperature	38 - 41° C
Bacteria	10^8 - 10^{10}/ml fluid
Gas Phase	Anaerobic, CO_2, CH_4
Solid Phase	Fibrous Mat
Liquid Phase	Volatile Fatty Acids (VFA) Ammonia Minerals Soluble Protein
VFA's	Acetate Propionate Butyrate

Also as a result of the continuous fermentation process, rumen temperature is slightly greater than the cow's and can contribute to helping maintain normal body temperature during cold weather or making the cow more uncomfortable during hot weather. The rumen has a specialized lining that contains many finger-like projections called papillae that absorb end products of fermentation, volatile fatty acids (VFA). The cow uses VFAs for energy (acetate, propionate, butyrate), fat synthesis (acetate, propionate, butyrate) or glucose (propionate exclusively) production. The rumen lining can be easily damaged by severe or prolonged declines in rumen pH, a result of excessive grain or insufficient fibre feeding.When the rumen is appropriately fed, it will contain a small gas cap, middle fibrous mat layer, and a lower liquid layer. The gas cap consists of carbon dioxide and methane, both end products of fermentation, which limits exposure of bacteria to oxygen. These gases must be regularly belched out to relive pressure, otherwise a potentially life-threatening condition termed bloat may occur.

The fibrous mat layer is composed of long dietary 'effective' fibre, which will help stimulate rumination and ruminal contractions. The tremendous number of bacteria found in the rumen are differentially distributed within the fibrous mat and liquid layers. Beside the type of raw materials microorganisms require for metabolism, reproductive rate also determines where the organism will be found in the rumen. Bacteria and protozoa that do not reproduce rapidly in relation to rate of passage through the rumen must attach to fibrous material if they are to remain in the rumen. When 'effective' fibre is not adequately provided, these microorganisms will be wiped out of the rumen and will result in abnormal fermentations and potentially digestive upsets and 'off-feed' situations.

The third ruminal chamber is the omasum, which is approximately the size of a basketball and located on the right side of the cow. The omasum is responsible for regulating particle passage rate from the rumen and water absorption from ingesta. Under normal rumen conditions, particles greater than 2 mm in size do not leave the rumen. Very little other information is known about this organ. When large fibre particles or whole corn kernels are found in the manure, this is a good indication of improper rumen function and should be further evaluated.

The abomasum, or fourth rumen chamber, is similar to our own stomach. Digestive enzymes and hydrochloric acid are secreted, which initiate breakdown of complex proteins and starches for further digestion in the small intestine. You may be more familiar with this organ from the problem associated with its displacement in early lactation. Left displaced abomasum (LDA) is a health disorder where the abomasum becomes atonic and distends with gas and passes under the rumen and becomes trapped on the left side of the cow. This condition has been associated with high grain and low fibre diets in early lactation and may also be related to subclinical hypocalcemia conditions.

Rumen Microbiology and Fermentation

Over 120 different species of microorganisms have been identified in the rumen. These organisms range from bacteria, the most abundant, to protozoa, fungi, and viruses. Although there is a wide variety of bacteria found in the rumen, they can be loosely grouped into five major categories in addition to protozoa. A basic understanding of the nutrient and environmental requirements of these different microbial groups is necessary to fully appreciate how feeding programmes may impact rumen health. Table lists substrates, requirements, and end products for these different microbial groups. One important concept to glean from this table is the observation that cellulolytic activity (*i.e.*, fibre fermentation) occurs only at higher pH levels.

A healthy rumen is one that has a balanced interaction between all groups of bacteria. In abnormal rumen environments, usually one group of bacteria has overwhelmed all other groups and dominates fermentation activity. For example, rumen acidosis is the result of feeding too much grain (sugars and starches), which allows starch digesters to overwhelm the rumen environment and eliminate cellulolytic activity. This is the crux of the problem in dairy cattle feeding, providing sufficient grain to support milk production without excessive amounts that can suppress fibre fermentation, milk fat test, and rumen activity.

Table. Characteristics of the different categories of microorganisms found in an anaerobic fermentation system.

Class of Organism	Primary Substrate	Specific Requirements	Primary Endproduct	pH Tolerance
Cellulolytic Bacteria (Fiber fermenting)	Cellulose Hemicellulose Pectins	Ammonia Iso-acids Cofactors	Acetate Succinate Formate, CO_2	Neutral 6.2-6.8
General Purpose Bacteria	Cellulose Starch	Ammonia Amino Acids	Propionate Succinate Butyrate Ammonia	Acid 5.5-6.6
Nonstructural CHO Bacteria	Starch Sugars	Amino Acids Ammonia	Propionate Lactate Butyrate Ammonia	Acid 5.0-6.6
Secondary Feeders	Succinate Lactate Fermentation Endproducts	Amino Acids	Ammonia Iso-acids Propionate	Neutral 6.2-6.8
Protozoa	Sugars Starch Bacteria	Amino Acids	Acetate Propionate Ammonia	Neutral 6.2-6.8
Methanogens	CO_2, H_2 Formate	Coenzyme M Ammonia	Methane	Neutral 6.2-6.8

Adapted from Chase, L.E. and C.J. Sniffen, Cornell University.

A number of factors can influence rumen fermentation efficiency. Most obvious is the role of diet on rumen pH and use of buffers to minimize the effect. In some countries ionophores are used to manipulate rumen fermentation patterns. Some recent research has shown that dietary mineral concentrations can influence rumen fermentation. Dietary supplementation of zinc (Zn) to maintain a rumen Zn concentration of 7 ppm was shown to inhibit ureolysis and increase molar proportion of propionate. High rumen

Zn concentrations (14 ppm) reduced fibre digestibility. Supplementing manganese (Mn) at 100 ppm in the rumen resulted in increased *in vitro* dry matter digestibility. More research is needed to fully determine the role of dietary minerals in altering rumen fermentation capacity.

Many rumen microbes are very sensitive to the presence of dietary polyunsaturated fats. Rumen microbes will attempt to reduce the metabolic toxicity of polyunsaturated fats by saturating double bonds through a process of biohydrogenation. Recent research has identified *trans*-10, *cis*-12 conjugated linoleic acid (CLA), a product of incomplete microbial biohydrogenation, to be associated with milk fat depression in dairy cows. The presence of trans-10 CLA inhibits or reduces mammary gland denovo fatty acid synthesis. The presence of large amounts of polyunsaturated fats in the rumen or small amounts with high grain feeding seem to promote the production of trans-10 CLA and produce milkfat depression syndrome.

NUTRIENT REQUIREMENTS: COW AND RUMEN

Required Nutrients

All living organisms require essential nutrients to support metabolic processes to keep them alive. General classification of required nutrients include: water, the most essential, energy, protein, minerals, and vitamins. Minerals can be further subdivided into macrominerals, microminerals based on the daily amounts (gm or mg) required. Vitamins are separated into fat or water soluble sources. Daily requirements for these essential nutrients are a function of the cow's body weight and physiologic state (*e.g.*, maintenance, growth, lactation, pregnancy) as modified by environmental conditions. Bacteria have similar requirements for maintenance and growth (*i.e.*, reproduction). Differences between the cow and microbes are seen in where they derive their nutrients. As a consequence of a pregastric fermentation system, consumed feeds are exposed to microbial fermentation before being available for digestion and absorption by the host. In many instances, this process enhances nutrient availability to the host animal. However, an opposite effect can also occur. Exposure to microbial fermentation can result in nutrient alteration to a form that is either no longer absorbed or has lost its biologic activity.

As an anaerobic fermentation system, the rumen is a highly reduced environment with an abundance of reducing equivalents looking for a place to go. Highly oxidized compounds are prime targets. Absorption of selenium (Se) occurs only in its highly oxidized forms as selenite (+4) or selenate (+6) via an intestinal sulfate transport system. Selenium absorption from selenite, is more efficient in non-ruminants than ruminants with a total retention of 77% and 29%, respectively. Microbial reduction of Se to its elemental (+0) or selenide (-2) forms renders it unavailable thus reducing Se absorption in ruminants. High concentrate diets are associated with increases in insoluble Se complexes, possibly due to changes in rumen pH, redox potential, microbial populations or some combination.

Table. Substances Which Supply Essential Nutrient Needs for the Cow and Rumen Microbial Population.

Nutrient	Cow	Bacteria
Energy	VFA's Glucose	Complex Carbohydrates Sugars, Starches, Amino Acids
Protein	Amino Acids Microbial Protein	Ammonia, Amino Acids, Peptides
Minerals	Dietary	Dietary
Vitamins	Dietary Bacterial	Dietary Synthesized

Rumen interactions can also result in a reduced availability of other minerals, namely copper (Cu) and magnesium (Mg). Best studied relative to copper availability are the interaction of molybdenum (Mo) and sulfate (SO_4) in the rumen. High Mo and SO_4 concentrations in the rumen allow microbial synthesis of thiomolybdates, which can chelate copper making it unavailable for absorption or utilization. Dietary Mg absorption efficiency is low and occurs in the rumen by a sodium (Na)-potassium (K) ATPase system. Many dietary factors, including excess K, nitrogen, Ca, sulfur and organic acids as well as deficient Mg and phosphorus levels, can predispose an animal to a form of hypomagnesemic tetany.

Another factor, a direct consequence of rumen microbial activity, also plays an important role in the pathogenesis of this syndrome. Grasses typically contain significant amounts (>1% of plant dry matter) of sucrose and trans-aconitate, an organic acid, during periods of rapid growth. In the rumen trans-aconitate is metabolized to acetate or reduced to tricarballyic acid. One of the most abundant rumen bacteria, *Selenomonas ruminantium,* is primarily responsible for the reduction of trans-aconitate. This organism also prefers to grow when sucrose is readily available as substrate. Tricarballylic acid is a very potent chelator of calcium (Ca), Mg and Zn inducing increased urinary excretion and reduced tissue status of these minerals.

Making more bugs - Microbial Protein

The cow derives a majority of her energy and protein from microbial end products. In other words, the more we make the bugs grow (reproduce), the less additional, more expensive feedstuffs we need to provide in the cow's diet. Why is production of microbes so important to a ruminant feeding programme? Microbes contain approximately 62% crude protein, which is 80% true protein and 80% digestible. This is considered high quality protein. Microbial protein production alone can support up to 25 kg of milk production. The first goal of a ruminant feeding programme should be to maximize microbial protein production and then secondly, meet additional cow's nutrient requirements over-and-above those not met by microbial fermentation end products. This type of feeding approach would theoretically be the most economical and efficient. So how do we get the rumen microbes to abundantly grow without disturbing the rumen ecosystem?

Table. Microbial Protein Synthesis Relative to Daily Protein Needs of the Cow.

Efficiency of Microbial Protein Synthesis	Daily Milk Yield		
	25 kg	35 kg	45 kg
gm N/kg OM digested	% of protein from microbes		
20	49	42	39
30	73	64	59
40	98	85	79

Bacteria require a number of essential nutrients for the synthesis of protein, similar to that of the cow. However unlike the cow, bacteria can use a greater variety of potential nitrogen sources to synthesize amino acids, the building blocks of proteins. In addition, bacteria can synthesize both essential and non-essential amino acids unlike the cow which needs to be supplied with preformed essential amino acids. Figure presents an overview of the processes required to synthesize microbial protein.

The microbial protein production is a function of rumen available substrates, primarily carbohydrates and nitrogen. If any of the required building blocks are in limited supply, microbial protein production will be determined by the availability of the most limiting substrate. Usually this is energy from carbohydrate fermentation. Energy production (generation of ATP) will be dependent upon the available carbohydrate source and its rate of degradation.

Ammonia (NH_3) may be provided from non-protein nitrogen sources, amino acids, peptides, or proteins where utilization of a nitrogen source is dependent upon the specific population of bacteria. For example, cellulolytic bacteria can only use NH_3 as their nitrogen source. Microbial protein production is more complex than just providing the necessary amounts of substrate in the diet. The rumen is a dynamic system that constantly has fermentation end products, liquid, bacteria, and particles being removed via digestion and passage through the rumen as well as new substrate added. So not only do we need to address concepts of total substrate requirements, availability of substrate relative to other substrates needs to be addressed. We must be able to

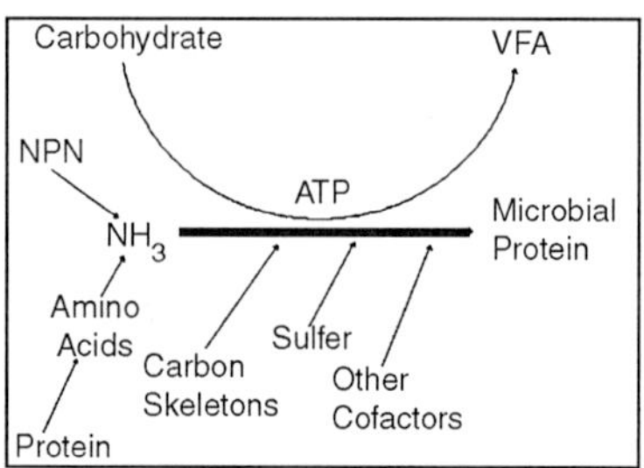

Metabolic processes involved in microbial protein synthesis.
NPN = non-protein nitrogen; NH_3 = ammonia.

Predict rate and extent of carbohydrate and protein degradation takes place in the rumen. This is the critical component of a dynamic modeling system for the rumen and requires more comprehensive and complex feed analysis procedures.

NEW CONCEPTS IN CARBOHYDRATE AND PROTEIN NUTRITION

Amino acids are supplied to the duodenum of ruminants by microbial protein synthesized in the rumen, undegraded dietary protein and endogenous protein. Microbial protein usually accounts for a substantial portion of the total amino acids entering the small intestine. Microbial protein is a high quality protein for the animal that is highly digestible in the small intestine. Because microbial growth rates affect amino acid supply to the ruminant animal, it is important to maximize microbial protein synthesis in the rumen.

Microbial protein synthesis in the rumen provides the majority of protein supplied to the small intestine of ruminants, accounting for 50 to 80% of total absorbable protein. The total amount of microbial protein flowing to the small intestine depends on nutrient availability and efficiency of utilization of these nutrients by ruminal bacteria. Ruminal degradation of protein from dietary feed ingredients is one of the most important factors influencing intestinal amino acid supply to ruminants. Proteolysis determines the availability of ammonia nitrogen, amino acids, peptides and branched-chain volatile fatty acids, which influence microbial growth rates in the rumen.

Rate and extent of ruminal proteolysis not only affect microbial protein synthesis but also the quantity and quality of undegraded dietary protein that reach the duodenum. Although microbial protein alone may be adequate for low producing ruminants, it can be inadequate for supporting higher levels of growth, wool or milk production. As animal production increases, additional protein must be provided from dietary protein that leaves the rumen undegraded to meet the animal's protein requirement. The use of high rumen undegradable proteins (RUP) in diets fed to ruminants with high protein requirements can improve the amino acid supply to the animal provided that enough degradable protein is included in the diet to maximize microbial protein. When formulating diets for ruminants, various criteria can be used to select the protein supplement including palatability, ruminal protein degradability, protein quality, intestinal absorption of amino acids, cost per unit of protein, availability and consistency of product, and impact on animal performance. Nutritional models for feeding protein to dairy cattle in the USA have evolved from basic crude protein (NRC, 1978) to more complex systems based on rumen degradable protein (RDP), RUP and intestinal digestion of RUP (NRC, 2001). The NRC (1989) recognized that intestinal digestion of protein supplements differs; however, empirical data were lacking and as a result, a constant value of 80% was used for all feeds. The main problem was the lack of reliable techniques for estimating intestinal digestion of proteins. With improved techniques, the NRC (2001) has assigned estimates of intestinal digestion to the RUP fraction of each feedstuff. These estimates were obtained from literature using the mobile bag technique and the three-step in situ/*in vitro* procedure of Calsamiglia and Stern (1995). Values used in the French Protein System were adopted for feeds with limited or no data.

Feeding proteins to ruminants that are resistant to microbial degradation in the rumen can provide a practical way to increase dietary protein and alter the amino acid profile of the protein reaching the small intestine for digestion and absorption. However, the effects of feeding high RUP on intestinal amino acid supply and animal performance have been inconsistent. Lack of response has been attributed to various factors including depression in ruminal microbial protein synthesis and factors related to quality of the dietary protein such as inadequate or overprotection of protein, reduced intestinal availability of amino acid or inherent amino acid limitations of the dietary protein. A considerable amount of variation among and within feeds in ruminal degradation and intestinal digestion of protein has been reported (Calsamiglia and Stern, 1995; Howie *et al.*, 1996; Yoon *et al.*, 1996). It is important that this variation be considered when determining the value of feeds as sources of protein for the ruminant animal.

Standardized chemical methods of feed analysis were developed over 150 years ago. The proximate analysis system that includes crude protein, crude fibre, ether extract and ash has been in use for over 100 years. However, this system is not adequate in characterizing feed composition relative to rumen and cow needs. Newer chemical and biological methods of feed analysis that better relate to nutritional function have been developed over the past 40 years and continue to be developed. The following describes the rationale and methods used to better characterize feed carbohydrates and protein fractions.

Understanding Carbohydrate Fractions

For the most part, carbohydrates (CHO) are in the diet to support rumen microbial populations. Various populations of microbes are capable of utilizing any CHO compound. In contrast, the cow has intestinal enzymes capable of only digesting sugars and starch. In fact, the cow like other mammals does not really have a CHO requirement as long as other glucose precursors are present in the diet. Therefore, dietary CHO need to be fractionated according to their capability of supporting rumen microbial growth and potential impact on rumen environment.

Carbohydrates are a tremendously diverse group of organic compounds and usually comprise more than 60% of the total diet in most dairy cattle rations. From a nutritional perspective, all CHO provide energy upon oxidation and depending upon their chemical structure may be a precursor for glucose (sugars, starches CHO) or fat (fibre CHO) synthesis. Due to the diversity and complexity in CHO structure, our ability to chemically characterize important nutritional fractions of CHO has been somewhat limited. The challenge for nutritionists has been to be able to adequately quantify different CHO fractions relative to their extent and rate of ruminal degradation or intestinal digestion as to determine their impact on animal performance.

Plant CHOs are primarily differentiated on the basis of their association to the cell wall. Carbohydrates that make up the cell wall are termed structural carbohydrates and are quantified as Neutral Detergent Fibre (NDF). Amount of NDF in a plant is determined by plant species and maturity. Structural CHOs have the property of being slowly fermented, if at all. Therefore, energy yield from these sources would be minimal and slow compared to non-fibre carbohydrates. Due to its slower fermentation rate and need for mastication, NDF is often associated with intake capacity in ruminants. A subset of NDF is acid detergent fibre (ADF), which quantifies the most slowly fermentable or non-fermentable portions of the cell wall. The ADF portion of feeds is often used to estimate digestibility or energy availability.

Non-fibre carbohydrates (NFC) are those compounds not associated with the cell wall, with the exception of neutral detergent soluble fibre (NDSF) compounds (pectic substances, fructosans, beta-glucans). Again, this is a very heterogeneous group of CHO that include organic acids, sugars, starch, and NDSF.

A subset of NFC is non-structural CHO (NSC), which are primarily sugars and starches that are very rapidly fermented in the rumen. In contrast to structural CHO, non-fibre CHO can rapidly provide large amounts of energy for microbial protein production or directly to the cow and their presence in a feed increases its digestibility. The diversity of compounds within this nutritionally defined group makes it difficult to directly measure amounts in a feed. Determination of non-fibre CHO is usually by difference, thus accumulating all errors in laboratory analyses for other feed fractions. Within this diverse group of compounds there is tremendous nutritional differences in rate and extent of fermentation or digestion. Ideally, we would like to be able to quantitatively subfractionate this group into organic acids, sugars, starch, and neutral detergent soluble fibre based on differences in nutritional responses. However, procedures to identify some of these non-fibre CHO fractions are only being developed.

Carbohydrate Terminology - The following is a summary list of commonly used CHO terms based on either chemical composition analysis or nutritionally-important fractions.

Crude Fibre (CF) - Original proximate analysis procedure of determining indigestible fibre (cell wall) content of feeds. This procedure incompletely accounts for total cell wall contents, due to limited recovery of hemicellulose and lignin, and thus underestimates total cell wall content of forages and roughages. Crude fibre content will always be less than NDF content and equal to or greater than ADF for a given feed.

Nitrogen Free Extract (NFE) - Proximate analysis procedure to determine readily available carbohydrate sources (sugars and starches). Determined by subtraction: (100 - CP - CF - Ash - EE). Due to the underestimate of fibre by crude fibre, NFE overestimates available carbohydrate, especially for forages.

Neutral Detergent Fibre (NDF) - Van Soest detergent methodology to recover all cell wall components excluding pectin. Contains primarily cellulose, hemicellulose and lignin as well as other resistant non-CHO substances. Correlated with dry matter intake in ruminant animals.

Acid Detergent Fibre (ADF) - Van Soest detergent methodology to recover the indigestible cell wall components. Contains cellulose and lignin as well as heat-damaged protein and other resistant plant compounds. Often associated with feed digestibility.

Non-fibre Carbohydrates (NFC) - Readily digestible (non-cell wall) carbohydrates as determined using the detergent system (organic acids, sugars, starches, soluble fibre). Similar to NFE determination, but always less than NFE for any given feed. Calculated by subtraction: (100 - (NDF + CP + EE + Ash)).

Nonstructural Carbohydrates (NSC) - those carbohydrates that can be digested by mammalian enzymes. Includes organic acids, sugars and starches. Newer chemical fractionation methods are allowing better determinations of starch in feedstuffs.

Lignin - Polyphenol compound, not a carbohydrate, but intimately associated with the plant cell wall carbohydrates. Totally unavailable portion of the cell wall. Lignin increases in amount within a plant structure with increasing plant maturity. Plant species differ in lignin amounts and its impact on carbohydrate availability. Lignin content (per cent DM) times 2.4 is used to estimate the amount of unavailable fibre in a feed.

Effective Fibre or NDF (eNDF, peNDF) - The portion of the total plant cell wall that is effective in increasing rumination and rumen motility. Effectiveness of NDF is based on particle size, degree of lignification, hydration, and density within classes of feeds. Effective fibre will increase rumen pH through its impact on rumination and saliva production.

Factors Affecting Rumen Carbohydrate Availability - A variety of factors beyond chemical composition of CHO can influence rate and extent of ruminal CHO degradation.

1. *Plant Maturity* - As a plant matures, there is an increase in the cell wall content thus diluting out the more digestible components (protein, sugars, minerals; Table). In addition as the plant matures, lignification of cell wall also increases making the cell wall less available for fermentation.

Table. Typical test value of alfalfa and grass hays harvested at various stages of plant maturity (all values on dry matter basis).

Type of Hay/Stage	CP %	ADF %	NDF %	TDN %
Alfalfa				
Pre-bloom	> 19	< 30	< 35	> 62
Early bloom	17-19	30-35	35-39	57 - 62
Mid bloom	13-16	36-41	41-47	51 - 56
Late bloom	< 13	> 41	> 48	< 51
Grass				
Prehead	17	< 29	< 55	> 54
Early head	12-17	30-35	56-61	47 - 54
Head	8-12	36-44	60-65	44 - 46
Post-head	< 8	> 45	> 65	< 44

Abbreviations: CP = crude protein; ADF = acid detergent fibre; NDF = neutral detergent fibre; TDN = total digestible nutrients.

2. *Environmental Conditions* - Rainfall, soil temperature, fertility, cloud cover, location, cutting strategies, etc. all can influence the availability of carbohydrates in the plant. Environmental light, temperature and their interaction have the greatest impact on plant growth. Increased temperature stimulates plant cell wall and lignification reducing plant digestibility. Light exposure will increase soluble CHO content making the plant more digestible.
3. *Processing* - Particle size reduction (grinding) increases surface area for available microbial attachment and degradation and is very beneficial in increasing cell wall digestion. Steam, extrusion, and popping will alter starch configuration to make it more available. Fermentation (ensiling) will make lesser available carbohydrates

more available. Heating can make soluble proteins insoluble and less rumen available.

4. *Plant Species* - Degree of lignification and distribution of lignin within the cell wall will affect rate of digestion of plant carbohydrates. Chemical structure of starch within a feed will dictate rate of fermentation or digestion. Corn starch (amylose) is less digestible than oat or barley starch (amylopectin), but if processed (ensiled, flaked, ground), it can become very available an a potential problem with ruminal acidosis.

Ruminal Nitrogen Metabolism

Ruminal Microbial Protein Synthesis The ultimate goal of proper rumen nutrition is to maximize microbial growth and the amount of RDP that is captured into rumen microbial cells. Maximizing the capture of degradable N not only improves the supply of AA to the small intestine, but also decreases N losses. The rumen is a complex environment inhabited by different microbial species, with each species having different nutrient requirements and metabolism. Therefore, considering the nutrient requirements of ruminal microorganisms is crucial to understanding N metabolism in the rumen as well as the factors that may modify it (Bach *et al.*, 2005).

Factors that affect microbial protein synthesis. Bacteria can utilize carbohydrates and proteins as energy sources. Carbohydrates are the main energy source for bacteria, although they can also be used as carbon skeletons for protein synthesis in combination with ammonia.

Ruminal microbial protein synthesis depends on supply of adequate amounts and type of carbohydrate as an energy source for the synthesis of peptide bonds. Readily fermentable carbohydrates such as starch or sugars are more effective than other sources such as cellulose in promoting microbial growth (Stern and Hoover, 1979). Several *in vitro* (Stern *et al.*, 1978; Henning *et al.*, 1991) and *in vivo* (Casper and Schingoethe, 1989; Cameron *et al.*, 1991) studies demonstrated that infusions of increasing amounts of readily fermentable carbohydrate decreased ammonia-N concentrations due to improved N uptake by ruminal microbes.

However, the optimum ratio of non-fibrous carbohydrates to ammonia-N has not yet been determined. In addition to the importance of the amounts of nutrient supply, the synchrony at which nutrients become available is also important. When rate of protein degradation exceeds the rate of carbohydrate fermentation, large quantities of N can be lost as ammonia, and conversely when the rate of carbohydrate fermentation exceeds protein degradation rate, microbial protein synthesis can decrease (Nocek and Russell, 1988). Whereas the concept of synchronous protein and energy supply has a solid theoretical basis, it is likely that in the complex ecosystem of mixed ruminal microorganisms when nutrient supply is synchronized for a specific subpopulation, it might not be synchronized for other populations. Therefore, average microbial efficiency remains fairly stable. Also, recycled N to the rumen may contribute to stabilize microbial growth even when N supply is not well synchronized (Bach *et al.*, 2005).

Ruminal bacteria are classified as cellulolytic and amylolytic based on their preferential use of energy. Russell *et al.* (1992) proposed a simplified model to describe energy and protein requirements of microbial subpopulations. Microbes that degrade structural carbohydrates (cellulolytic) have low maintenance requirements, grow slowly, and use ammonia-N as their main N source, whereas microorganisms that degrade non-structural carbohydrates (amylolytic) have higher maintenance requirements, grow rapidly, and use ammonia, peptides, and amino acids (AA) as N sources (Russell *et al.*, 1992). However, bacterial growth has been shown to increase with addition of AA and (or) peptides in cellulolytic and amylolytic bacteria (Maeng and Baldwin, 1976; Argyle and Baldwin, 1989; Kernick, 1991). Similarly, fibre digestion was reported to increase with the supply of AA (Griswold *et al.*, 1996; Carro and Miller, 1999) and peptides (Cruz Soto *et al.*, 1994) to pure cellulolytic bacteria.

Atasoglu *et al.* (2001) demonstrated with pure cultures of cellulolytic bacteria, that the incorporation of ammonia-N into microbial cell-N decreased as the proportion of AA increased in the medium, suggesting that cellulolytic bacteria would use AA if available. Similar findings were reported with increasing concentrations of peptides, although Atasoglu *et al.* (2001) reported a greater preference of cellulolytic bacteria for incorporating AA-N compared with peptide-N into their cell-N. However, at typical ruminal peptide and AA concentrations, about 80% of the cell-N is derived from ammonia-N. Addition of branched-chain AA that will ferment to branched-chain VFA, and addition of peptides to ruminal fluid has increased fibre digestion, microbial protein production and growth efficiencies (Russell and Sniffen, 1984; Thomsen, 1985). The increase in microbial growth observed with addition of AA and (or) peptides, may be due to direct incorporation of AA into microbial protein and (or) to increased availability of carbon skeletons from AA deamination, which can be used for energy production or as carbon skeletons for new microbial AA.

Importance of microbial protein to ruminants. The theoretical contribution of microbial protein to the total protein requirement of the lactating dairy cow, calculated at three efficiencies of microbial protein synthesis, is presented in Table. Contribution of microbial protein to total protein requirement was determined using NRC values (2001) for a 680-kg lactating dairy cow producing 25, 35, or 45 kg/d of 4% FCM. At these three milk yields, microbial protein would contribute 51, 49 and 48%, respectively, of the total protein required by the cow when microbial synthesis in the rumen is 30 g of N/kg of organic matter truly digested (OMTD). When milk yield is 45 kg/d, the contribution of microbial protein to total protein required by the cow would increase from 32 to 63% as efficiency of microbial protein synthesis increases from 20 to 40 g of N/kg of OMTD. Stern and Hoover (1979) reviewed the literature and reported that approximately 30 g of N were synthesized per kilogram of OMD in the rumen; values ranged from 10 to 50 g. Efficiencies of microbial protein synthesis used in Table fall into the range of reported values; therefore, calculated contributions of microbial protein clearly depicted the importance of optimizing microbial protein synthesis in the rumen of high producing dairy cows. In addition, these calculations demonstrate that, as milk yield increases, a substantial quantity of RUP from protein supplements must leave the rumen to meet the protein requirement of the cow.

Contribution of microbial protein to growing beef cattle calculated at three efficiencies of microbial protein synthesis is presented in Table. Contribution of microbial protein to total protein requirement was determined using NRC values (1996) for a 250 kg growing beef animal with average daily gains (ADG) of 300, 1200 and 1700 g/day. At these three ADG, microbial protein would contribute 95, 57 and 40%, respectively, of the total protein required by the animal when microbial synthesis in the rumen is 25 g of N/kg of organic matter truly digested (OMTD). Lower efficiencies of microbial protein synthesis, ranging from 20 to 30 g N/kg OMTD, were used because beef cattle are generally less efficient, most probably due to a high population of amylolytic bacteria. Microbial protein contribution to total protein requirement demonstrated a more dramatic difference in beef cattle compared with dairy cattle. This response is probably due to the lack of increase in dry matter intake in beef cattle as ADG increases. However, in both situations, it is clear that substantial quantities of RUP from protein supplements must be incorporated into the diet of high producing ruminants.

Estimates of rumen undegradable protein. Under ideal circumstances, the ruminant nutritionist would prefer to use "book values" for RUP of various protein supplements, such as those found in the NRC (2001) publication. However, ruminal protein degradation is a complex process influenced by various factors such as solubility, protein structure, microbial proteolytic activity, microbial access to the protein and ruminal retention time of dietary protein. There is an inherent variability in reported values for RUP of feeds that is related to conditions under which experiments are conducted; however, part of the variation can also be attributed to differences in quality of protein supplements.

Understanding Protein Fractions

Dietary crude protein (N% x 6.25) is not a very useful measure of what potentially happens with consumed nitrogen relative to utilization by the rumen or cow. Crude protein can be chemically separated into fractions on the basis of rumen degradability and solubility and have nutritional relevance to meeting the protein needs of the rumen and cow. Proteins that are rumen degradable would be able to provide nitrogen for microbial protein production. Rumen solubility suggests that the protein source would be more rapidly available. For example, urea, a non-protein nitrogen source, is 100% soluble and degradable and therefore would very rapidly provide ammonia for microbial protein production. Rumen insoluble and very slowly degraded is that dietary protein that can bypass to the abomasum to be digested enzymatically by the cow. The amount of unavailable dietary protein is measured as the amount of nitrogen found in the ADF fraction.

In meeting our goal of maximum microbial protein production, we need to match carbohydrate and protein rates of degradation. This allows for somewhat equivalent amounts of rumen energy and nitrogen availability promoting efficient microbial protein yield and overall dietary protein incorporation. If we can reduce the amount of protein used in the ration and yet maintain or improve milk yield, the dairy cow becomes much more efficient, profitable and environmentally friendly! Protein Terminology - the following is a summary list of commonly used terms relative to protein nutrition of ruminant animals.

- *Crude Protein (CP)*: Total nitrogen per cent of a feed times 6.25 factor. Crude protein analysis does not differentiate between true protein and non-protein nitrogen sources.
- *Digestible Protein (DP)*: Amount of protein absorbed from the intestine from a feed source. Digested protein in the ruminant animal comes from microbial protein and undegraded dietary protein.
- *Soluble Protein (Sol CP, SIP)*: Dietary protein that readily goes into solution in rumen fluid. This is a rapidly available source of NH_3 for the rumen bugs. Composed of NPN and soluble true proteins.
- *Rumen Degradable Protein (RDP, DIP)*: Dietary protein that can be fermented in the rumen by microbes and contribute to the rumen ammonia pool. Rate of passage of digesta through the rumen will determine the extent of which RDP will be degraded.
- *Rumen Undegradable Protein (RUP, UIP), Escape Protein, Bypass Protein:* Protein which is either very slowly degraded in the rumen or which is unavailable to rumen fermentation. This protein is available for digestion in the abomasum to provide amino acids for the cow.
- *Metabolizable Protein (MP)*: Dietary protein (feed and microbial) that has been digested and is available for absorption and utilization.
- *Non-protein Nitrogen (NPN):* Nitrogenous compounds that do not contain linkages of amino acids. Potential protein source for rumen microbes but the utilization is dependent upon dietary CHO availability. Sources include urea, ammonia, biuret and amino acids.
- *Unavailable Protein, Bound Protein (ADIN, ADF-N)*: Protein that is unavailable to rumen degradation or abomasum digestion. This is

the amount of nitrogen present in the ADF residue and is a measure of protein bound to fibre due to heat damage. It is very indigestible and reduces feedstuff quality.

Intestinal Protein Digestion

Methods for Measuring Intestinal Protein Digestion The total amount of protein available for absorption from the small intestine depends on the flow of microbial and dietary protein to the duodenum and their respective intestinal digestibilities.

Digestion of protein that leaves the rumen starts in the abomasum with acid-pepsin digestion and is completed in the small intestine with pancreatic and intestinal proteases. The NRC (1989) recognized that intestinal digestion of protein supplements may differ; however, empirical data were lacking, and as a result, a constant value of 80% was used for all feeds. The main problem was the lack of reliable techniques for estimating intestinal digestion of proteins (Stern *et al.*, 1997).

In vivo estimation of intestinal protein digestion involves expensive and labor-intensive experiments and requires the use of surgically prepared animals. Apparent digestion of protein is calculated as the disappearance of CP or amino acids between the duodenum and ileum, which is subject to considerable error associated with digesta sampling, use of digesta flow rate markers, and inherent animal variation. Therefore, several alternative procedures for estimation of intestinal protein digestion in ruminants have been developed.

MAKING A DYNAMIC RUMEN MODEL

Over the past few decades much research has been devoted to quantification of total dietary nutrient needs of dairy cows to support all levels of performance. Many computerized feeding programmes a based on comparison between nutrients delivered in the diet and defined daily nutrient requirements. However, this approach is not sensitive to dynamic changes in feed intake and other factors that impact feed degradability and availability, which ultimately impact the amount of metabolizable nutrients available to the cow to support production.

Nutrient availability from a feed either by microbial fermentation or intestinal digestion is not constant. Competition between inherent feed properties of degradability or digestibility and rate of passage through the digestive tract can explain the amount of a given nutrient pool will be available. This interaction of rates is mathematically described in equations. This concept is the central component of any dynamic rumen model such as the Cornell Net Carbohydrate and Protein System (CNCPS).

- Amount Degraded = Pool size x $K_d / [K_d + K_p]$
- Amount Passed = Pool size x $K_p / [K_d + K_p]$

How much of a given carbohydrate or protein fraction within a feed is degraded will depend upon how rapid the rate of degradation is compared to rate of passage. Rate of passage through the rumen depends upon the fraction. Forages or fibre are slowly passed between 4 and 8 per cent/hr whereas the liquid fraction passes more quickly, between 12 and 20 per cent/hr. If a feed fraction has a rate of degradation that greatly exceeds rate of passage, then most or all of the fraction will be degraded in the rumen. If rate of degradation is slower than rate of passage, then only a small fraction will be degraded in the rumen.

The dynamic components to this system are those CHO or protein fractions where their rate of degradation is nearly equal to the rate of passage. For these fractions, *i.e.*, fermentable fibre CHO and insoluble degradable protein, slight changes in either rate of passage or degradation can greatly influence how much of the fraction will be degraded in the rumen. The following describes a variety of factors that can influence ruminal degradation and passage rates.

Rate of Degradation (K_d)

Degradation of a feed ingredient is inherently related to the chemical and physical properties of all component compounds. Relative to CHO feeds, degradation (ruminal or intestinal) of sugars is extremely rapid compared to starch or fibre. Within starches there is some range of degradation. Fibre is more slowly fermented, while some fibre is not fermented at all. Similarly with proteins, soluble proteins are rapidly degraded while others are more slowly or not at all. It is important then to have an accurate chemical analysis of feed ingredients to determine how much of each of these different fractions comprise a given feed ingredient.

Beyond chemical and physical properties of a feed, degradability can be modified by feed processing, grinding, and rumen conditions. Heating, grinding, ensiling or some combination will increase rumen degradability of feed CHO starch and fibre fractions. For example, corn starch is very crystalline in nature and hydrophobic, thus being fairly resistant to fermentation without processing. If corn grain is ensiled and ground starch availability will be increased dramatically. This explains why coarsely ground corn may be seen in manure compared to ground high moisture corn causing acidosis. Ensiling will increase protein availability to the rumen through increasing solubility. Heat treatment of protein makes the insoluble degradable fraction less rumen degradable. Proteolytic activity in the rumen will dictate extent of rumen protein breakdown of degradable fraction. A decline in rumen pH will greatly reduce fibre fermentation and may reduce proteolytic activity.

Rate of Passage (K_p)

Rate of ruminal passage is calculated for forages and concentrates. Rate of passage will always be slower for forages compared to concentrates. Dry matter intake, forage concentration in the diet and body weight are the primary determinants of rate of passage. Rate of passage increases with increasing dry matter intake and dietary forage concentration. Rate of passage declines slightly with increasing body weight. For each feed an adjustment factor to rate of passage is calculated based on effective NDF of the feed. Passage rate will be reduced with higher eNDF content of feeds.

Putting the Concepts Together

When designing a ruminant feeding programme, concepts of rates need to be addressed in attempting to make sure ruminal availability of energy and nitrogen are coordinated in order to achieve maximal microbial protein production. The goal is to achieve some level of synchrony between ruminal energy and nitrogen sources in order to maximize microbial protein yield. If energy and nitrogen sources are not synchronized, then loss of available energy or nitrogen and reduced microbial activity will occur. Asynchronous energy and nitrogen sources will result in reduced milk production and efficiency and elevated blood urea nitrogen values. In this scenario diseases such as urea toxicity, grain overload or other rumen dysfunctions may occur. If energy and nitrogen sources are synchronized, we maintain maximal microbial protein synthesis. In this scenario not only will the amount of additional dietary protein be reduced, but the cows will have increased dry matter intake and remain healthier, all contributing to increased milk producing efficiency. How is this synchronization achieved?

Intuitively, synchronization of ruminal energy and nitrogen sources is achieved by utilizing various CHO and protein feed ingredients that have differing degradability properties. Use of a single protein (canola meal or soybean meal) or energy source (corn grain) does not give one this flexibility. Using ensiled feeds as the sole forage programme unbalances protein to the more readily degradable fractions. One can alter degrability properties with grinding, heating or some other processing procedure as previously discussed. This is where the strength of the dynamic formulation programmes comes into place. One can use such software to appropriately balance feed ingredients to best meet ruminal needs. In addition, one can use milk or blood urea nitrogen as a diagnostic aid in evaluating the balanced achieved between dietary CHO and nitrogen sources.

Another more simplified way of achieving relative synchrony to the rumen environment is through the feeding programme. Feeding a total mixed ration (TMR) is the single best method of approximating rumen nutrient synchrony. The needed balance between CHO and protein fractions is achieved with multiple meal feeding. This effect accounts for the observed 5-8% improvement in milk production when shifting a conventional to a TMR feeding programme. More slowly degraded CHO fractions can trap readily available nitrogen at a later meal. When concentrates are fed in the milking parlor, this balance is lost. In a conventional feeding system where feed ingredients are fed separately, one then needs to more closely account for differences in CHO and protein availability within feeds. Thus, sequence of feeding can have an impact on milk production response. Forages should be fed prior to concentrate meals to maintain slower rate of passage. Large concentrate meals are best divided and fed 3 to 4 times or more per day to limit the effect of acidosis.

FEEDING RECOMMENDATIONS FOR DIETARY CARBOHYDRATES AND PROTEIN

The preceding discussion has focused on understanding the rumen system and presented concepts on how to better feed the microbial populations to improve milk production efficiency. To move into this new era of dairy cattle feeding, more comprehensive methods of feed composition analysis are needed. With a better understanding of feed composition, we can apply concepts of rates of degradation and passage to a dynamic rumen system to achieve improved feeding practices that will not only improve milk production efficiency, but decrease wasted dietary nitrogen from entering the environment. The following tables provide some practical feeding guidelines in applying these concepts to ration formulation.

NDF and Dry Matter Intake

As previously described, NDF has been shown to influence dry matter intake in ruminant animals. Mertens had shown optimum intake of NDF as a per cent of body weight (%BW) was 1.2 " 0.1 %BW. Neutral detergent fibre intake capacity is influenced by physiologic state and age. Cows can consume more or less dietary NDF, however, there will be negative consequences on production or performance as a result of reduced intake.

Non-fibre Carbohydrates

The NFC portion of the diet is primarily responsible for rapid ruminal fermentation in support of milk production. Milk production will be severely compromised in the face of low dietary NFC content. Conversely, excess dietary NFC content may result in ruminal dysfunction, namely subclinical or clinical acidosis. The balancing act with NFC is to provide sufficient amounts to support maximal microbial growth, yet not providing excess amounts that will reduce ruminal pH and compromise fermentation activity. Table presents suggested dietary concentrations for total NFC and NSC (sugar and starch) components over differing physiologic states.

Protein Fractions

From a dietary protein formulation perspective, nitrogen needs of the rumen should be met first, then metabolizable protein requirements for the cow should be addressed. Dietary crude protein is a poor method by which to formulate diets in this manner. However, one needs some sophisticated dynamic models to adequately predict dietary metabolizable protein delivery.

If one formulates diets to the specific protein fractions, the amount of metabolizable protein should be reasonably adequate. Again, one can only truly predict the degradability and undegradablility of protein sources with a dynamic modeling system, which accounts for dry matter intake and ingredient interactions on availability and contribution to either microbial protein yield or digestible undegraded protein.

Table. Suggested Dietary Concentrations for Various Protein Fractions in Diets for Dairy Cattle.

	Crude Protein	Soluble CP	RDP[1]	RUP[1]
Dry Cows	% DM	% of CP	% of CP	% of CP
Far Off	12 - 13	35	70 - 72	28 - 30
Close-up	14 - 15	30	63 - 65	35 - 37
Lactating Cows				
Fresh	19	30	60 - 62	38 - 40
22-80 DIM	18	31	64	36
81-200 DIM	16	32	66	34
>200 DIM	14	34	68	32

Abbreviations: RDP = rumen degradable protein; RUP = rumen undegradable protein.

In formulating for dietary protein, rumen nitrogen needs must be considered first. Ruminal degrable protein should be between 10.5 and 11% of dry matter to support microbial nitrogen needs. One can never separate rumen nitrogen from CHO needs. To address this interaction between dietary CHO and protein in the rumen, one should formulate for a NFC:RDP ratio between 3.0 - 3.5 to 1.

This suggests as NFC content goes up, then more degradable protein can be assimilated. Rumen protein needs should be met with forage and plant source proteins. If additional bypass protein is necessary, then heat-treated plant proteins or animal proteins (if allowable) can be used to meet the cow's metabolizable protein needs. Use of rumen protected amino acids should be left to only fine tune a diet if necessary, given their cost.

MORINGA FODDER IN RUMINANT NUTRITION

Within the framework of market-oriented system improvement programme research at the International Trypanotolerance Centre, Banjul, the need to improve the feed resource base as an ameliorative move across the farming systems is overwhelming. This is principally a response to one of the most crucial needs of the livestock sub-sector – nutrition. From time immemorial, the groundnut residue has been the traditional feed resource of choice especially in urban areas. These crop residues are hauled into peri-urban animal depots where they are used on a zero-grazing basis. However, due to economic and agronomic reasons, during the last decade, it has become necessary to investigate the potentials of alternative feed resources. Consequently, the International Trypanotolerance Centre, Banjul screened a cohort of more than thirty five feed resources that can be used by urban ruminants as basal and/or supplement diets.

The Moringa oleifera variously known as drumstick tree, etc has been systematically investigated during the last three years at ITC. The *Moringa* plant is well known for its enormous biomass production and it promises to be the plant of the future in ruminant animal supplementation strategy. Although not completely strange in the West African biosphere, this grossly underexploited plant has a lot to offer as a food and fodder resource in the sub-region. Considering the agro-ecological characteristics of the sub-region in general and The Gambia in particular, the plant needs to be further investigated and modalities of integration into the farming system carefully conducted on the merits of the respective locations.

The plant *Moringa oleifera* is very popular (locally known as 'never die') in the Senegambia region where it exists principally in scattered uncultivated forms. It is mainly consumed in various forms as food in the region although a few useful extra-culinary attributes have been suggested. Reports from other tropical countries around the world, especially in Latin America and the pilot trials conducted in The Gambia (Adediran and Akinbamijo, unpublished) have shown very encouraging biomass yields. More recently, under high density cultivation, biomass yields in excess of 15 tonnes DM/ha in a 60-day growing cycle has been obtained at the International Trypanotolerance Centre, Banjul. This volume of high quality biomass is simply overwhelming given the context of the semi arid ecologies in The Gambia. This is of particular interest to animal nutrition where dietary protein sources are becoming increasingly expensive and difficult to access.

Given this magnitude of potentially useful biomass, there has been no systematic attempt to exploit *Moringa* either in terms of its agronomic attributes or its nutritional values. The preliminary investigations were therefore aimed at the systematic investigation of the nutritive evaluation and the modalities of establishing *Moringa oleifera* in the various farming systems in West Africa From the growing body of information on the plant, it is common knowledge that it is endowed with exceptional physicochemical properties with the possibilities of high biomass yield described above obtained at ITC.

In addition, it is known to be devoid of any known anti nutritive factors and insignificant tannin contents. The plant is known contain at least 25% protein. Compared with other conventional ruminant feedstuffs in The Gambia, it has a very high biological value and considerable potential for adoption as food for humans as well as ruminant fodder resource

SYSTEMATIC INVESTIGATIONS IN THE GAMBIA

Since the beginning of the century, a new research strategy on the *Moringa* initiative was launched in The Gambia. This initiative received a lot of logistical support from the Institute for Animal Production in the Tropics and subtropics of the University of Hohenheim and the Church World Service, Dakar and the National Nutrition Agency of The Gambia.

With the compelling attributes demonstrated in The Gambia, there was every reason to pursue the *Moringa* research plan in the context of the Senegambian faming system. Within this system, *Moringa* will readily fill both human and animal nutrition gaps. The ultimate beneficiaries therefore appear to be poor agro-pastoralists that need enormous food and feed support to keep their lives and those of their animals (upon which their livelihoods depend) in a sustainable manner. The nutritive value of *Moringa oleifera* is currently undergoing vigourous tests on-station and on-farm in parallel in the context of the West African farming systems and ecology.

AGRONOMY

The *Moringa* plant is cultivated in a high density mode with three approaches adopted in the evaluation of *Moringa oleifera* in the West African farming system context. The first step was the agronomic assessment of biomass yield under diverse production and harvesting regimes. Following the encouraging results obtained from a pilot study on *Moringa oleifera* establishment and yield at the International Trypanotolerance Centre, Banjul in 2001 with a planting density of 15cmX15cm, a logical next step was the determination of the optimum cutting height and intervals that would provide the best quality of fodder as well as retain the biomass yield at a relatively high level.

In theory, we were to determine the maximum biomass yield without significant negative impact on the re-growth potential of the plant under reasonable conditions that can support re-growth. Three cutting heights – 20, 30 and 45 cm were tested in three cutting interval cycles of 50, 100 and 150 days after planting. Fodder was assessed for biomass yields, fibres and digestible nutrients.

Biomass Evaluation

Biomass yield obtained by estimate dry matter content of the materials recovered in a 1 sq meter quadrat indicated the possibility of obtaining up to 20 tonnes DM/ha in a 50 day growing cycle. However, for the purposes of easy communication to the ultimate beneficiaries, it is recommended that the cutting cycle be rounded to once in two months rather than every fifty days. With a cutting interval of only 60 days before it is used as fodder, the nutritive value of the fodder is very high with very low chances of lignification. It must be stated however that optimum growing conditions such as continuous irrigation, high planting density and a fertilizer regime of 50 kg/ha/month are imperative before such high biomass yields can be obtained.

In vitro Assays

In order to improve our understanding of the nutritive value of *Moringa*, it was compared with other *in vitro*-simulated combination diets in a series of laboratory-based studies using the Hohenheim Gas Test method to quantify corresponding rumen gas production to *Moringa*-based diets. Emerging trends indicate that the *Moringa* option compared with groundnut cake, has a significantly higher gas production potential of up to 36 ml gas compared with 28 ml of VFAs at 40% inclusion in the diet. Using the Hohenheim Gat Test procedure, we have been able to improve our understanding on the comparative advantages of *Moringa* on groundnut hay.

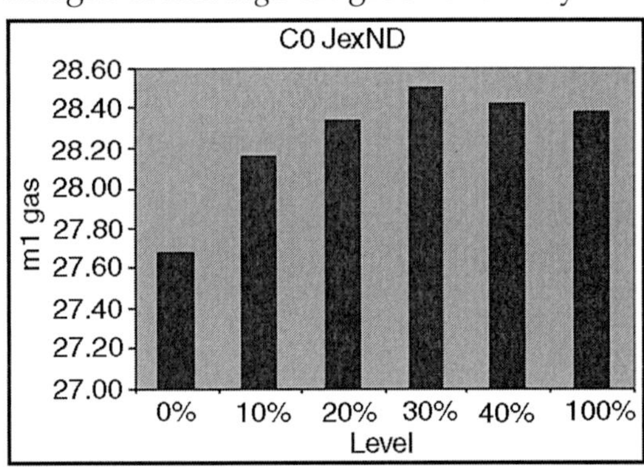

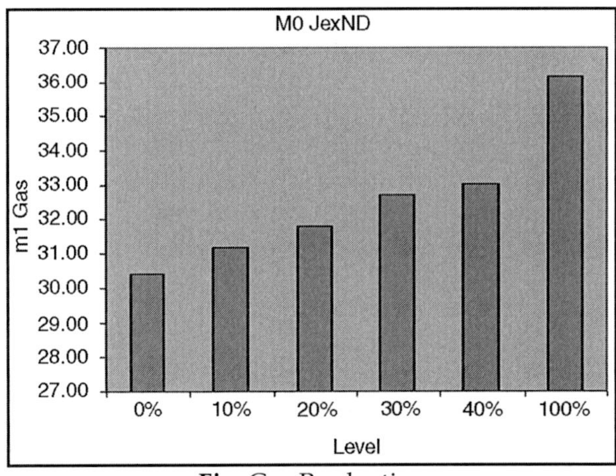

Fig. Gas Production

Ruminant Nutrition: High density nutrient supplementation strategy

Moringa oleifera was investigated as an alternative to conventional concentrates in defining an appropriate supplementation strategy in a dairy production system using two classical methods:

- *In vivo* digestibility
- *In vivo* validation trial

In addition, animal response on a trial diet using *Moringa* to supplement a groundnut hay- based diet has shown that there is no significant difference in performance when compared with animals supplemented with groundnut cake based concentrate.

In a short term animal response study set up to validate the *in vitro* results using growing crossbred animals, in quantitative terms, animals offered *Moringa* leaves as supplements did attain a higher growth rate of 440 g/d compared to their counterparts offered conventional locally available concentrates (1:1 mixture of GNC and rice bran) that grew at the rate of 385 g/d although this difference did not attain statistical significance (Fig 2). As expected, the supplemented animals offered either *Moringa* or concentrate out-performed their counterparts offered only groundnut hay and gained only 274 g/d. The growth rate of the control animals was significantly lower than those of their supplemented counterparts.

LESSONS LEARNT

The present results confirms the enormous potentials of *Moringa* as a supplement for ruminants especially in highly intensive production systems such as the peri-urban dairy systems and as ameliorative dry season feeding strategy in the traditional extensive system of husbandry.

Challenges

- Processing of biomass to avoid post harvest losses especially during the rainy season

On-farm production will be a seasonal affair unless for exceptional cases where there are farmers that can afford to irrigate their *Moringa* fields. Our experiences in The Gambia with on-farm intensive production are that there is a high likelihood of excessive biomass yield during the rainy season. In order to optimize the exploitation of *Moringa*, the issue of fodder preservation becomes imperative in order to avoid post harvest losses. Efforts on how to address the perishability of the fodder is a current subject of research at the International Trypanotolerance Centre, Banjul.

- *Storage for use during the dry season:* As on-farm production is geared towards food and feed, appropriate storage and methods and facilities have to be in place prior to harvesting. It is common knowledge that *Moringa* is a nutrient-rich supplement in terms of minerals and vitamins essential for vital physiological processes in humans and livestock. In order to avoid post harvest depreciation in the nutritive values of the nitrogen free extracts, there is need to provide adequate long-term storage facilities available to *Moringa* growers.

- *Year round production in intensive systems:* In peri-urban dairy systems, where feeding is required to cater for the different phases of the dam cycle, appropriate cow nutrition becomes necessary.
- Land availability and communal ownership of high density *Moringa* fields

The introduction of high density *Moringa* cultivation is still a novel approach to most of the end users. Although the plant itself is not new, the end users have never needed to have dedicated plots for growing *Moringa* in large quantities. As land tenure remains a big issue in areas where *Moringa* is most needed – especially as periurban feed resource, it is important that policy instruments in urban areas are formulated with feed resource concerns in perspective. In the rural areas where *Moringa* can greatly improve human and animal nutrition on community basis, there is need to adopt community forest ownership strategies to enhance the adoption and establishment of communal *Moringa* fields.

EFFECTS OF TANNINS ON RUMEN FUNCTION

It has been believed for some considerable period that tannin above 5% can become a serious anti-nutritional factor in plant materials fed to ruminants (McLeod, 1974). Barry (1983) and his colleagues have demonstrated with *Lotus pedunculatus* that the ideal concentration of condensed tannins in this forage legume is between 2–4% of the diet dry matter, at which level they bind with the dietary proteins during mastication and appear to protect the protein from microbial attack in the rumen. On the other hand, other forages with tannins have anti-nutritional effects in the rumen and reduce N retention (Waghorn & Shelton 1995). If the protein-tannin complex dissociates under acid conditions then the protein can be digested in the lower gut. At higher levels (5– 9%) tannins become highly detrimental (Barry, 1983) as they reduce digestibility of fibre in the rumen (Reed *et al.* 1985) by inhibiting the activity of bacteria (Chesson *et al.* 1982) and anaerobic fungi (Akin & Rigsby, 1985) high levels also lead to reduced intake (Merton & Ehle, 1984); above 9% tannins may become lethal to an animal that has no other feed (Kumar, 1983).

Sheep have been shown to adapt slowly to tannins in a diet of *Acacia* leaves, suggesting that there are rumen organisms that in some way detoxify their effects (Reed *et al.* 1985). Recent studies have demonstrated a bacterium in the rumen of goats and sheep capable of growth at high tannin levels in the rumen (Brooker, J., personal communication). Thus a little tannin has been usually accepted as being able to protect protein of forages and allow a higher efficiency of feed utilization by the animal. However, recent results throw some doubt on this. Tannins may indirectly effect rumen function by reducing rumen ammonia levels through decreased protein degradation in the rumen. If rumen ammonia levels decrease below 80 mg N/l then fibre digestibility may be depressed and digestibility is reduced well below 10 mg N/l (Leng *et al.* 1993). Whenever tannins are present in forages there may be a need to supplement ruminants with a non-protein nitrogen source such as urea or chicken manure. Conversely, tannins in feed may increase detrimental effects on rumen function when the basal diet is low in protein.

Tannin in Plant Foliage

Condensed tannins are present in only some plant species. In general, shrub and tree foliages are likely to be higher in tannins than pasture plants, and leguminous forages from the tropics are generally higher in tannin than those from the temperate countries. The level of tannins within a species has been found to vary considerably depending on a number of factors. For example in New Zealand, *Lotus pedunculatus* grown on fertile, high moisture soils has about one third the condensed tannin content of *Lotus* grown on hill country under water stress (Barry, T., personal communication).

The literature in this area is often confusing as reported levels of tannins often seem to be higher than can be explained and these values often are rejected by reviewers as possibly due to errors of analysis. For example, *Prosopis* leaves have been reported to have tannin levels of 2.2% of the dry matter (Sehgal, 1984) but in 15 individual trees Joshi *et al.* (1985) reported levels ranging from 10.6 to 25.3% which suggested to some authors that there may have been major analytical difficulties. Certainly the level of tannins in the foliage of trees is highly variable and depends on environmental stress (fertility, soil-water relationships, insect attack etc.). Newer information also points to analytical flaws in preparation of leaf material for analysis and drying reduces the measured tannin levels over fresh material. New leaves often have higher tannin content than older leaves (Vaithiyonathan & Singh, 1989) and in South Africa the tannin content of*Acacia* grazed by Kudu increased with grazing pressure (van Hoven, 1991).

Tannins at 14–16% are present in the bark of many trees (Dalziel, 1948) indicating a very large pool in the tree that can possibly be mobilized. In general it could be expected that the green bark of new growth would contain less tannin than the brown bark, and the leaves and petals less than bark.

Tannin Mobilization

There is a small amount of literature indicating that the tannin content of some tree foliages may be controlled in some way and can be elevated at times of high risk of defoliation (*i.e.*, by insect attack, cutting and harvesting, or grazing).

An investigation in South Africa of the death of a number of Kudu on a wild life farm has led to knowledge that may have major implication for management of some fodder trees (van Hoven, 1985). A number of Kudu (member of the deer family) died after grazing on a small area of *Acacia* trees (van Hoven, 1991). Subsequent studies have shown that in the wild these Kudu would approach such woodlands and after grazing on the trees on the periphery move quickly to trees well separated from those recently grazed. However, because of the enclosure these animals were forced to consume more of the foliage from one woodland. Subsequent studies implicated tannin in the death of the animals and this led to a study of tannin in the tree foliage. Tannin levels rose sharply over a 15 minute to 1 hour period, not only in the trees grazed by deer, but in the trees adjacent to those that were being damaged.

Nomads in India, grazing camels on *Prosopis juliflora* on the roadside carefully explained the need to move their camels to fresh trees after a short time of grazing in the one area (Leng, R.A. personal observation). Although not yet tested, this fits the idea of some trees responding to grazing by increasing the content of anti-palatability secondary plant compounds. The implication is that there is a pool of tannin which can be readily mobilized by activation of specific enzymes sensitive to air borne materials released from damaged foliage. This obviously has been an important survival mechanism for trees on the savannahs. Also it has major implications for the use of browse trees in pasture/tree associations for cattle production. However, responses of some tree foliages to such treatment is far from clear and a number of trees do not respond to damage in any way. This area requires experimental work in a number of situations. The effect of simulated grazing on tannin content of a number of tree leaves is shown in Table. The information on such responses in a number of trees needs to be compiled.

The Implications of Tannin Build-up in Foliage

Tannin buildup in foliage has enormous implications for the use of fodder trees. It may partially explain some of the contradictory nature of much of the data on the efficiency of use of fodder trees, and the common rejection of trees for fodder purposes in some areas and not in others. The quality of harvested foliage will always depend on where it is produced, how it is harvested, the stage of plant growth, the climate, the effects of insect damage and the stocking rate. Where trees respond to damage, it is possible that occasional grazing by animals in a plantation may effect tannin levels in the leaves for up to 4 days.

As tannins appear to be both anti-nutritional and perhaps nutritionally beneficial it is important to determine what factors influence tannin levels in foliage; whether these factors can be controlled (*e.g.*, strategic fertilizer or water application) and whether harvesting techniques and stocking rates can be developed to optimize or minimize the tannin levels. If it is too complex to develop such techniques then it is obvious that high tannin levels may be acceptable so long as they are diluted by other feed resources that have sufficient protein to bind free tannins beneficially but this necessitates a cut and carry system for tree fodders.

Tannins in the Rumen

The fate of tannins following ingestion by ruminants depends on the type of tannin. Most tannins form complexes with protein in the plant material, in saliva or the rumen contents. Hydrolyzable tannins hydrolyze in gastric acidity beyond the rumen, releasing protein, amino acids and small units of phenolics that probably pass to the urine. At high levels of tannin intake, both mucoproteins and the epithelial cell lining of the digestive tract are affected. This alters the integrity of the gut wall causing problems of gastritis, slowed propulsion of feeds and constipation (Kumar & Singh, 1984). Only under exceptional circumstances will herbivores consume large quantities of forages containing tannins, although recent studies from Zimbabwe suggests that the intake of browse can be increased by feeding very small quantities of polyethylene glycol (Duncan, 1994). However, from the point of view of this publication, fodder trees are considered only as potential supplements to poor quality forage and not a basal feed reserve.

Drying promotes the combination of tannin and plant protein before ingestion which may result in different responses by ruminants on forage diets to supplements of dry or fresh tree foliage. Proanthocyanidin tannins are held in special organs in the leaves to prevent their interference with the plant's own metabolic apparatus, and this factor favours tannin-tolerant animal browsers that have tannin-binding proteins in their saliva (Austin *et al.* 1989). In these instances the salivary factor binds the tannins and spares valuable forage protein with a higher concentration of essential amino acids.

Attempts to use tannins to promote rumen escape ("bypass") of feed protein have had limited success. Fresh forage may be ingested before tannin-protein complexes can oxidatively cross-link, as in the case of dry feeds, although drying has been shown to have favourable effects on productivity. Recently, Waghorn & Shelton, (1995) showed that, in sheep fed freshly cut rye grass, the addition of a third of the diet as *Lotus pedunculatus* (which supplied 1.8% condensed tannin in the total diet) resulted in the digestibility of the protein being significantly decreased from 78% to 65%. The inclusion of *Lotus* forage also lowered dry matter digestibility by 3–7%, which was almost all accounted for by the lowered protein digestibility.

Apparently the level of protein availability in the animal from plant and other sources was unaffected, as wool growth and liveweight gain were not different in groups given *Lotus* forage compared with those on pure rye grass. These experiments suggest that condensed tannins overprotect protein and that the precipitation of protein-tannin complexes protect the micro-organisms in the rumen from the detrimental effects of condensed tannins. The presence of tannins in forages stimulate salivary flow in animals. A number of studies have shown an apparent increase in microbial protein leaving the rumen after feeding moderate levels of tannin (Beever & Siddons, 1986), but the level of increase is not convincing, considering the technology used to make the measurements. Nitrogen balance is apparently improved in animals that are fed low levels of tannins, although digestibility of forage fibre may be lowered. The effect of condensed tannins overall appears to be to make more amino acids available in the intestine. Whether this is the result of increased microbial growth efficiency or increased dietary protein availability is unclear.

Plant Compounds in Fodder Trees

Many foliages have chemicals that appear to be produced for the purpose of deterring invasion or consumption of their leaves by microbes, insects and herbivorous animals. Whilst tannins are the best known of these, there is a long list of secondary plant compounds.

Cyanide, nitrate, fluoroacetate, cyanogenic glycosides, saponins, oxalates, mimosine and various sterols are but a few. Quite recently, saponin concentrations have been implicated in low productivity in young cattle on young growth of signal grass (*Brachiaria*) (Lowe, S., personal communication). These may or may not be modified in the rumen by microbial action. The primary compound or its breakdown products in the rumen may be toxic or of no nutritional consequence. Some secondary plant compounds are actively detoxified in the liver.

One of the best known secondary plant compound in tree leaves is mimosine in *Leucaena,* which is degraded to the toxic compound 3-hydroxy-4-pyridone by normal rumen organisms and which is usually further degraded by another microbe where animals have evolved in grazing situations containing *Leucaena.* This microbe is not present in ruminants that have been isolated from *Leucaena* and inoculation is necessary to prevent toxicity when *Leucaena* becomes a high proportion of the total diet (Jones & Megarrity, 1986).

The main reason for the foregoing discussion is to emphasize that these secondary plant compounds should be taken into consideration. However, the toxic compounds often only become of significance nutritionally when the plant assumes a high proportion of the diet. Although some effects of the toxic compounds may persist when these are used to supplement the diets, the beneficial effects of a high protein forage often override their effects. There are, however, important exceptions particularly in the case of fluoroacetate which is an lethal compound that claims the lives of large numbers of cattle on pasture lands with *Acacia georginae,* even though it is of relatively low digestibility. Recently a rumen microbe has been genetically modified to carry and express a gene that encodes for an enzyme that hydrolyzes fluoroacetate (Gregg *et al.* 1994).

Comparisons of Tree Foliages and Multinutrient Blocks

Tree foliages have been given high prominence as protein supplements for ruminants fed low protein forages. However, seldom is it known whether escape protein *per se* is the valuable component of the foliage or whether the protein of the tree foliage is largely providing ammonia (from protein degradation), minerals, or all three. The rôle that tree foliages play in ruminant nutrition determines the required rate of supplementation, so it is extremely important to know what this rôle is.

Tree foliages low in tannins or other secondary plant compounds that might bind protein, are probably degraded rapidly and, at times, completely. In this way they provide ammonia and volatile fatty acids in the rumen. The rate of microbial growth on protein, however, is approximately half that on carbohydrate, so P/E ratios are lower when protein is degraded in the rumen in comparison with carbohydrates. On the other hand, if foliage protein is bound by condensed tannins, microbial degradation of leaf protein in the rumen will be prevented or slowed.

This will allow particles high in protein to move to the lower digestive tract where some of the condensed tannin complexed with protein may be hydrolyzed, which then allows the protein to be digested. Condensed tannin under acid or alkaline conditions of the intestines may be split to sugars and organic acids, mostly gallic acids, releasing protein and amino acids that are digestible in the lower gut. However, lowered N retention in animals fed some tanniniferous forages suggest that much bound protein, is unavailable for digestion in the gut (Mangan, 1988; Kumar & Singh, 1990; Waghorn & Shelton, 1995). Provided some protein remains soluble and can provide ammonia in the rumen and perhaps escape protein, some of the detrimental effects would be reduced. In this case the protein level relative to tannins would be the most critical factor.

Few research studies have defined the rôle required of tree foliages where their feeding objective is as supplements to low digestibility forage. The extent to which tree foliage protein is degraded in or escapes the rumen is extremely important. If the tree foliage protein is totally degraded then it provides only ammonia and minerals for microbial growth (but both these may be more easily and economically provided from other sources such as MUMB, chicken manure or litter from broiler production).

Tree foliages, on the other hand, often have higher digestibilities than the pasture forage available and thus provide a more energy dense feed, allowing a higher total feed intake. In a situation where the fermentable N and minerals can be provided more economically, say, as MUMB then the use of such tree foliages is counter-indicated. The potential value of the foliage may only be realized by harvesting and processing so that it contains a higher percentage of bypass protein.

11

Nutrition and Food Requirements of Animal

NUTRITIONAL REQUIREMENTS OF GRAZING ANIMALS

The nutrients required by animals are energy, protein, vitamins and minerals. The concept of requirements is generally seen as the amounts necessary to support "normal" metabolic activity. That is, the animal's requirements are thought to be met when it gives evidence of normal health and vigour, normal rate of growth, normal reproduction and/or normal lactation levels. Obviously, "normal" is not identical in all members of the same species at all times so these requirements should be seen as a set of ranges.

Nutrients as limiting factors, while an important concept, should not be thought of as a rigid one-to-one relationship. Generally, nutrients are utilized in the hierarchical order of maintenance, reproduction, lactation and storage. However, across a population of animals, reproduction and lactation can occur when the diet does not provide the "required" levels for these functions. Within that same population, a certain proportion of animals can even reproduce or lactate at nutrient levels well below maintenance "requirements." Despite the absence of rigour, the concepts of nutrient requirements and priority of use are fundamental to an understanding of animal nutrition and management.

COMPARATIVE PROTEIN AND VITAMIN NUTRITION

Proteins are large molecular compounds comprised of approximately 20 individual amino acids bonded together in linear, coiled or branching chain forms. The relative number of each of these amino acids and the sequence in which they are bonded together determine the character of the particular protein in the tissue (muscle, hair, hoof, enzyme, etc.). Protein in the diet must be broken down to the individual amino acids within the gastrointestinal tract and absorbed as such since the large protein molecule cannot be transported through the intestinal wall.

These absorbed amino acids are then used to resynthesize proteins that fit the needs of the animal. Some of the amino acids can be formed within the tissue from materials such as other amino acids that are present in excess. On the other hand, some of the amino acids must be absorbed from the gastrointestinal tract preformed and are referred to as essential amino acids. If absorbed amounts of essential amino acids meet or exceed the animal's physiological requirements, protein synthesis can proceed at a normal rate. If one or more are not absorbed in sufficient amounts, tissue protein synthesis is restricted and the associated maintenance or production function is impeded.

Vitamins are "cofactors" or catalysts in metabolic reactions, in that they do not appear in the products of reactions, but must be present for reactions to occur. All vitamins or their precursors must be absorbed from the digestive tract as they cannot be synthesized by mammalian tissue. If vitamins are not absorbed in adequate amounts, metabolic activity is restricted. Protein and vitamin nutrition are both influenced by microbial fermentation and its location within the digestive tract. Pathways of protein synthesis in microorganisms are similar to those of mammalian tissue except that amino acid requirements are much less specific. The microorganisms, as a mixed population, have no absolute amino acid requirements. Ammonia, derived from most nitrogen-containing compounds, including urea, can be used in the synthesis of "microbial protein". Likewise, most vitamins are synthesized by populations of microorganisms except for vitamin A, D, and E.

Ruminant animals are insulated against essential amino acid and most vitamin deficiencies because these compounds are synthesized by symbiotic microbial populations in the rumen and subsequently presented for hydrolytic digestion in the gastric-intestinal region. Once microbial protein passes from the rumen to the gastric-intestinal region, it is hydrolyzed to the individual amino acids which are absorbed for use at the tissue level. Therefore, ruminants can survive on a protein-free diet as long as the diet contains a form of nitrogen to yield ammonia under anaerobic fermentation (Virtanen 1968). Additional insulation against protein deficiency is conferred by ammonia nitrogen recycling (Weston and Hogan 1967). However, over the longer term, a base supply of amino acids in the form of dietary protein may be necessary for maximal fibre digestion and ruminal protein synthesis (Petersen *et al.* 1985). Vitamins synthesized by the microorganisms in the rumen are likewise digested in the lower tract.

The essential amino acids necessary to achieve and sustain maximum production, defined as rapid growth, successful reproduction and heavy lactation in domestic ruminants, cannot be met solely through microbial protein synthesis (Burroughs *et al.* 1975). Microbial growth is limited by the maximum level of fermentation which can be supported by a given diet (substrate). Obviously, complete fermentation of a substrate in the rumen can yield only a finite amount of microbial protein. Even at maximum fermentation, microbial synthesis is unable to provide sufficient quantities of amino acids to fully satisfy the physiologic requirements for maximum productivity (genetic potential) of some particular animals in a highly productive state (*e.g.*, rapidly growing). Maximum productivity can be achieved only by the addition of escape protein, with a favourable amino acid profile, to augment microbial protein production.

DIGESTION AND FLOW DYNAMICS IN RUMINANTS

The chemical components of the diets of ruminants can be separated into two structural fractions of nutritional significance. The first, cell contents (neutral detergent solubles; NDS), are those substances found inside plant cells. These organic molecules are soluble and so are readily digestible in the intestine. These substances also tend to be rapidly and extensively fermented before reaching the gastric-intestinal region. The second fraction, cell wall components (neutral detergent fibre; NDF) are digested more slowly and less completely. Digestion of the cell wall fraction is performed almost exclusively by microbial hydrolysis and fermentation. Volatile fatty acids produced during fermentation are absorbed through the rumen wall and subsequently metabolized for use at the tissue level as energy. Products of fermentation not absorbed through the rumen wall, including microbial cells, pass to the lower tract together with unfermented dietary residues. These modified (or synthesized) and original dietary and endogenous fractions are exposed to hydrolytic digestion in the gastric-intestinal region.

The total amount and quality of nutrients derived from a grazing animal's diet is determined by the type and amount of forage consumed and the proportioning of the material among five possible fates:

- Fate 1. Degraded to products absorbed directly from the compartment (VFAs);
- Fate 2. Modified during fermentation in the rumen and subsequently digested in the lower tract (microbial protein);
- Fate 3. Escape fermentation in the rumen and undergo hydrolytic digestion in the gastric-intestinal region (bypass or escape protein);
- Fate 4. Either modified by or escape fermentation in the rumen and/ or hydrolytic digestion in the gastric- intestinal region to be fermented and absorbed in the colon/cecum;
- Fate 5. Bypass or escape digestion completely and excreted in the feces.

Some plant components such as cellulose are of greater nutritive value to the animal if they remain in the rumen over an extended period and are either degraded to absorbable end products (*e.g.*, VFAs; Fate 1), or their fermentation contributes to the formation of other substances (*e.g.*, microbial cells) that are subsequently digested (Fate 2). Otherwise such components are destined to either minimal nutrient yield from colon/cecal fermentation (Fate 4) or be excreted (Fate 5). Other components such as non-fibre bound protein and soluble sugars have greater net value if they escape ruminal fermentation and undergo hydrolytic digestion in the abomasum and small intestine (Fate 3) because respiration losses associated with anaerobic microbial fermentation are avoided. Therein lies the reason ruminants can survive, and are indeed productive, on fibrous forage diets, but are comparatively inefficient, compared with the chicken or pig, at converting feeds high in soluble carbohydrates and protein to animal products.

Thus, inherent species differences in gastrointestinal flow dynamics ultimately influence which species are adapted to particular components of the vegetation on rangeland. Cattle and bison, which have a relatively large capacity rumen compartment in relation to both body size and nutrient requirements (Demment and Van Soest 1985), also have a long rumen retention time (RT). These anatomical factors permit cattle to extract a large amount of nutritional value from fibrous materials, often in amounts adequate to satisfy all their nutrient requirements.

Conversely, small ruminants, *e.g.*, sheep and goats which possess a relatively small rumen compartment in relation to body size and nutrient requirements cannot extract comparable levels of nutrients from the same fibrous forages. Even though nutrient requirements are greater per unit body weight in small ruminants, rumen capacity is significantly less, retention time significantly shorter and flow rate significantly faster than in large ruminants. Hence for relatively equivalent intake levels, fibrous diets are of less nutritional value to small ruminants.

FORAGING STRATEGIES OF RUMINANTS

Smaller ruminants have evolved two strategies to overcome the metabolic. The first strategy is reduced RT (Van Soest 1982) which allows a slight shift in the site of digestion of the highly digestible components out of rumen fermentation (Fate 1 and/or Fate 2) and into the gastric-intestinal region (Fate 3) thereby decreasing respiration losses associated with fermentation. Also, the shorter RT is associated with a greater level of intake and a slightly depressed fibre digestibility.

Taken together this results in a greater level of intake, a slightly lower digestibility compared to larger ruminants, but an opportunity to equal or exceed total digested nutrient intake (Huston 1978). This strategy is important in survival but is seldom effective in allowing the small ruminants to match the productivity of large ruminants when both are limited to high fibre diets. The second strategy is to consume a high quality diet which necessitates a greater degree of discrimination in diet selection. Size and prehensile agility of the lips, teeth and tongue ultimately determine an animal's ability to selectively consume plant species, individual plants on offer within a species, and even discrete plant parts, all from a heterogeneous assemblage of plant biomass. Significant differences in the morphological structure of mouth parts exist in pre-gastric fermenters and post-gastric fermenters which reflect the types of forages consumed. Generally, increased pliability of the lips and manipulative capacity of the tongue denote greater levels of selectivity.

Range herbivores have been variously classified into as many as six classes based upon the types of foods eaten (Langer 1984). T modified form of the system described by Hofmann and Stewart (1972) applied to ruminants. Bulk/roughage grazers (cattle, bison, cape buffalo, etc.) graze comparatively indiscriminately on the herbaceous fraction of vegetation by wrapping their tongue around individual clumps of plant growth and, with a short jerking motion of the head, break the clump loose then draw it into their mouths. Once in the mouth, the material is wetted with salivary secretions, chewed slightly, formed into a cylindrical "bolus" with the teeth and tongue, then swallowed. Later, when the animal is at rest, swallowed material is regurgitated, chewed extensively, then reswallowed (rumination).

Concentrate selectors (white-tail deer, mule deer, dik-dik, etc.) characteristically have pliable and often split lips, soft muzzles and agile tongues. Hence, these animals can select plants or plant parts high in cell contents (protein and other soluble fractions; NDS) and low in cell wall (cellulose and fibrous fractions; NDF). Bite sizes are smaller and more discrete, even consisting of single leaves, leaf tips, fruits, seeds or fallen mast. Intermediate feeders are a diverse group characterized by dietary plasticity not found in either bulk/roughage feeders or concentrate selectors. Diet is characterized by variety and frequent compositional changes. The domestic sheep is classified as an intermediate feeder, but its diet often approximates the bulk/roughage group. The goat is a true intermediate feeder, and its diet selections clearly overlap the entire array of forages.

Such a classification of the feeding behavior of grazing animals is useful to better understand species adaptability to specific forage conditions but should not lead the reader to believe these are rigid relationships because "crossover" in feeding habits regularly occurs. Especially within sympatric ruminant populations, all species select diets from an array of available plant materials which vary in space and time. Availability is the first and most important determinant of what a grazing animal consumes. When the opportunity is presented for selection among types, species and morphological parts of plants, ruminant populations regularly exhibit "preferences" in the materials selected. This ability to discriminate between available materials is sufficiently pronounced that in vegetatively productive periods the diets of ruminant species grazing in common are almost completely different Conversely, during periods when the amount and diversity of forage are limited, dietary overlap between sympatric species is very high.

COMPARATIVE NUTRITIONAL PHYSIOLOGY OF HERBIVORES

The distinction of ruminants relative to their adaptability to forage-based animal production systems, stems from three characteristics unique to this group of animals. First, by virtue of the evolution of a pre-gastric fermentation chamber, ruminants can more effectively utilize structural carbohydrates (NDF) than either non-ruminants or post-gastric fermenters of comparable size. Increased retention time under conditions of anaerobic fermentation leads to more complete digestion and utilization of forage. It must again be noted that ruminant species vary widely both in RT and the extent of fermentive degradation of forage components.

Secondly, whereas non-ruminants depend on preformed amino acids and vitamins in their diets, ruminants are comparatively free of these requirements. Simple forms of dietary or endogenous nitrogen (ammonia releasing compounds *i.e.*, urea, proteins, amino acids, etc.) can be used by ruminants in the microbial synthesis of protein which subsequently is digested in the gastric-intestinal region. This adaptation is further enhanced by the ability to recycle urea via salivary and ruminal mucosal secretions. Microbial protein generally fulfills the minimal amino acid requirements of ruminants for maintenance and moderate levels of production. Genetically possible levels of production in animals in stages of high productivity cannot be achieved without the addition of escape protein to increase the supply of essential amino acids.

Lastly, dietary overlap of sympatric animal species can be very high or low depending upon forage diversity and availability, environmental conditions and management. The net effect of these three physiological and behavioral characteristics is that ruminants, as a group, are well adapted to production systems on rangeland.

FOOD REQUIREMENTS OF ANIMAL

VITAMINS

Vitamins are organic molecules that are necessary for normal metabolism in animals, but either are not synthesized in the body or are synthesized in inadequate quantities. Consequently, vitamins must be obtained from the diet. Most vitamins function as coenzymes or cofactors. Deficiency states are recognized for all vitamins, and in many cases, excessive intake also leads to disease.

The fact that human beings were subject to various diseases as a result of deficient diet has been long known. Even the early Egyptians knew that night blindness could be cured by eating the livers of domestic animals. Scurvy occurred among the crusaders of the thirteenth century. In 1520 the Austrian surgeon, Kramer, pointed out that scurvy could be cured by oranges, limes, or lemons, and in 1795 lemon juice was introduced in the English navy as a preventive of the disease. In 1882 the Japanese admiral Takaki (or Takagi) was able to cause a marked decrease in the number of cases of beriberi in the Japanese navy by improving the diet of the sailors. Vitamin research really began with the discovery that animals could be used in experimentation with deficiency diseases. The first clear-cut experimental work is that of Eijkman. He found that if chickens were fed polished rice they developed a disease, called polyneuritis, very similar to beriberi. This disease could then be cured by feeding the birds the polishings which had been removed from the rice. Following this important experiment, progress was for a time slow, and most workers in the field of nutrition did not realise its significance.

Hopkins was one of the exceptions. As far back as 1906 he noted that no animal could live on a diet of pure protein, fat, and carbohydrate, and that obscure dietary factors were responsible for such diseases as scurvy and rickets. It was not until 1915 that it was realised that more than one type of vitamin was involved in nutrition. In that year McCollum and Davis showed that two types of unknown substances were necessary for the normal nutrition of the rat. At first these were called "fat-soluble A" and "water-soluble B," but it was not long before the term vitamin was accepted throughout the world, and fatsoluble A and water-soluble B became known as vitamin A and vitamin B. Soon there came to be a growing alphabet of vitamins. Then followed a period in which vitamin B began to be subdivided into B1, B2, etc.

At the present time it is customary to speak of the vitamins A or the vitamins of the B complex. As we come to know the true chemical composition of the individual vitamins, there is an increasing tendency to call them by their correct chemical names instead of by letters. Thus vitamin C is now commonly referred to as ascorbic acid. There is no clear-cut definition of vitamins, no sure way of deciding whether a substance important in nutrition is to be regarded as a vitamin or not. Vitamins are organic substances capable in small concentration of affecting the health or influencing the rate of growth of organisms. There are numerous substances that act in this way. Some of them are regarded as vitamins, others not, or there may be differences of opinion.

Thus it was noted in the last chapter that certain unsaturated fats were necessary for the growth and well-being of rats. These fats can be regarded as a fat requirement or they can be called vitamin F. Similarly, various insects require cholesterol. This may be regarded simply as a sterol requirement, or the cholesterol may be considered a vitamin in the same sense that calciferol (vitamin D) is a vitamin. The fact that vitamins are capable of exerting a powerful influence when present in extremely low concentrations is an indication that they act in some way as catalysts. Certainly this is true of some vitamins. It is by no means an easy task to present an up-to-date summary of our information concerning vitamins. New papers are being published at a rapid rate, and it is well-nigh impossible for anyone not a specialist in the field to keep abreast of them. For the student, perhaps the easiest way to become informed is to read some book on the subject or some recent treatise on physiological chemistry. The earlier studies of vitamins were essentially practical. Men suffered from diseases due to vitamin deficiency. It was important to discover which foods contained the necessary vitamins, and in what amount. Lacking a chemical knowledge of the vitamins it was essential to develop biological methods of assay. Thus, it was found that a certain amount of vitamin was necessary to cure pigeons of polyneuritis or to permit normal growth in rats. On the basis of such studies on rats, pigeons, and guinea-pigs, various types of vitamin units were established, and the vitamin content of all sorts of foods was determined.

To raise colonies of rats and guinea pigs is rather an expensive and time-consuming business. Fortunately, the presence of vitamins and the amount of a certain vitamin contained in a given sample of food can be tested with organisms much easier and less expensive to raise than mammals and birds. Insects require many of the known vitamins and they might be used as test animals. But much more suitable are various bacteria and fungi. They can be raised in test tubes in a minimum of space. Obviously one must first determine the exact vitamin requirements of a given type of bacterium or fungus. The assay of vitamins by the use of bacteria is now well established practice and there is a large literature in the field. This has been reviewed by Snell. In 1941, Beadle and Tatum by x-ray treatment were able to produce mutant strains of the fungus Neurospora which are highly useful for vitamin assay. The usual type of Neurospora requires only the vitamin biotin, but by the use of x-rays and ultraviolet radiation within a few years mutants were obtained which had a specific requirement for every one of the B vitamins then known with the single exception of folic acid.

Similar mutants have also been obtained in bacteria. The chemical composition of most vitamins is now known. This makes possible direct chemical analysis. Often such analysis is aided by spectroscopic study. Thus the A vitamins give a blue colour with antimony trichloride and this colour is the basis for spectroscopic tests. Some vitamins give characteristic absorption bands in the ultraviolet. Likewise, some vitamins are fluorescent, and the fluorescent colour can be tested spectroscopically.

Thus in addition to ordinary chemical assay methods, spectroscopic tests are of great value. The subject is discussed at length in volume 1 of György *Vitamin Methods*. In many cases a vitamin may be replaced by other chemical substances related to it chemically. Thus various substances related to ascorbic acid (vitamin C) behave like ascorbic acid in preventing scurvy. There are many substances with vitamin D or vitamin K activity. The fact that a substance has a chemical structure similar to that of a vitamin does not necessarily indicate that the substance will act like the vitamin. Often, indeed, a substance which resembles a particular vitamin in its chemical composition is an antagonist to it and tends to prevent its action. So, for example, Dicumarol is chemically close to vitamin K; and yet whereas feeding of vitamin K favours blood clotting in higher animals, Dicumarol prevents clotting. Similarly, various analogues of pyridoxine (a B_6 vitamin) act to prevent its action. Other cases of anti-vitamins could be cited. Studies of antivitamin action may give clues as to the mechanism of the action of the vitamin they antagonize. The subject is also of considerable interest to the theoretical pharmacologist. More and more cases are being discovered in which the harmful action of a particular drug is due to its chemical affinity to some substance important for the vital process.

In some cases, the need for a vitamin is lessened by feeding a substance which can be converted into it. In some organisms, tryptophan can be converted into nicotinic acid. Thus, chicks do not require nicotinic acid if fed enough tryptophan. Moreover, isotope studies show that tryptophan can be converted into nicotinic acid. Another instance in which a necessary vitamin can be replaced by an amino acid has been described for the lactic acid bacteria, Streptococcus faecalis and Lactobacillus casei. Certain strains of these species require vitamin B_6, but they can grow without this vitamin in the presence of *d*alanine (the dextro-rotatory, naturally occurring form of the amino acid). In this instance, however, the amino acid is not the precursor of the vitamin, but apparently the vitamin is necessary in order to produce the essential amino acid.

These cases, interesting in themselves, indicate also that the need of a particular organism for a given vitamin may to some extent depend on the diet. As a result of our knowledge concerning the chemical nature of the vitamins, many of them can be synthesized and are actually being manufactured on a large scale. We also have much knowledge concerning the specific vitamin requirements of many higher animals, of the isolated roots of green plants, of bacteria, fungi and protozoa. More and more information is becoming available concerning the vitamin needs of various insects. Vitamin studies are relatively simple for terrestrial forms-mammals, birds, insects. Aquatic forms are difficult material.

Perhaps if fish could be grown in sterilized water and kept sterile, correct data on their vitamin requirements could be obtained. With ordinary methods of attack, one can never be certain of the contribution that bacteria and other microorganisms are making to the vitamin content of the food taken in by the fish. Even in terrestrial animals, microorganisms may play a part in providing vitamins.

The bacteria in the intestinal tract can and do manufacture vitamins. In experiments with rats, it is sometimes necessary to prevent the animals from eating their feces, for such fecal material may be a source of vitamins. In many instances a mammal may not need a particular vitamin as food because it is manufactured by the bacteria of its intestine. By giving the animal antibiotics such as sulfa drugs and thus eliminating or reducing its intestinal bacterial flora, it is possible to investigate the need for a vitamin such as folic acid. The nutritional requirements of various types of animals may depend largely on the parasites or symbionts they harbour. Many species of insects, especially among the Hemiptera, have a constant association with symbiotic fungi. Even protozoa may contain symbionts or parasites within their single-celled bodies. Most of the work on vitamins has been stimulated by the practical needs of medicine and industry. The physician and the drug manufacturer are not ordinarily interested in the vitamin requirements of lower organisms, unless these organisms can conveniently be used in vitamin assay. Thus, as might be expected, there are vast gaps in our knowledge of the vitamin requirements of many large groups of animals. We know almost nothing of the vitamin needs of metazoan invertebrates other than insects.

And we know very little about the vitamins necessary for fish, amphibia and reptiles. A wide field is open for such study, and many facts of interest may well be discovered. Some vitamins are required by practically all living organisms from bacteria to man; others are needed by only some forms and not others. There has been much interest in the avitaminoses, that is to say the diseases caused by the lack of specific vitamins. Our knowledge in this field is almost entirely confined to higher animals. Thus we know that when the vitamins D are omitted from the diet, a child will develop rickets and so will a pig or a rabbit, a dog or a chicken. Accordingly, the vitamins D have been called the antirickets vitamins. But animals without bones such as insects or protozoa may also require vitamin D. From a knowledge of the reasons why this is so, we may be able to obtain additional information as to the way that vitamins function.

Much as is known about the vitamins, their chemistry, their occurrence in various foods, the diseases that follow their absence from the diet, there is still almost complete ignorance as to the exact nature of their action. A beginning has been made in the recognition of the fact that some vitamins such as riboflavin, nicotinic acid, and pantothenic acid form essential parts of some enzyme systems.

We know also the end results of many types of vitamin deficiency. But to say that a vitamin is necessary for growth or that it prevents anemia tells us very little as to what it really does or why it is necessary. If we were to know how certain vitamins influence growth or prevent cell deterioration in the epithelial or nervous system, we would have basic information of great importance. Such information would be useful not only in the better understanding of the vitamins but also in the elucidation of various basic vital processes. The final interpretation of vitamin action must lie in an understanding of how the vitamins affect the cell protoplasm.

It is this aspect of the subject that will be stressed in the following discussion of individual vitamins. The general physiologist is not primarily interested in the relation of vitamins to human welfare, the foods a man should eat, or the benefits to be derived from adding this or that vitamin to the diet. Such information can be obtained in books on physiological chemistry, nutrition, or medicine. In discussing the individual vitamins, we shall not proceed in alphabetical order. This order has no other basis than the chronology of discovery. We shall consider first the fat-soluble and then the water-soluble vitamins. In general, especially in the case of the water-soluble vitamins, true chemical names have tended to supplant letters and subscripts. But in some instances, several chemical substances can and normally do supply a given vitamin need. In these instances the alphabetical name is sometimes more convenient.

VITAMIN A

Vitamin A and its derivatives (retinoids) are essential components in vision; they contribute to pattern formation during development and exert multiple effects on cell differentiation. It has been known for 70 y that the key step in vitamin A biosynthesis is the oxidative cleavage of a carotenoid with provitamin A activity. While a detailed biochemical characterization of the respective enzymes could be achieved in cell-free homogenates, their molecular nature has remained elusive for a long time. Recent research led to the identification of genes encoding two different types of carotene oxygenases from animal species. The molecular cloning of these different types of animal carotene oxygenases establishes the existence of a family of carotenoid metabolizing enzymes in animals heretofore described in plants. With these tools in hands, old questions in vitamin A research can be definitively addressed on the molecular levels contributing to a mechanistic understanding of the regulation of vitamin A homeostasis or tissue specificity of vitamin A formation, with impact on animal physiology and human health.

Vitamin A is a yellow viscous oil. Chemically, it is related to carotene, the common yellow pigment of plants. The generally accepted formula follows:

H_2C CH_2

C   H H CH_3 H H H CH_3 H

H_2C C–C–C–C—C–C–C–C—C CH_2OH

H_2C C–CH_2

C

H H

When carotene is eaten, it is converted to vitamin A. Because of this, it is sometimes called a provitamin. Another provitamin of vitamin A is kitol, a dihydric alcohol found in whale liver oil. The conversion of carotene to vitamin A occurs in the intestines of higher animals. The vitamin is then stored in the liver. The livers of polar bears and Arctic foxes are so rich in vitamin A that they are toxic. The toxicity is an example of what is known as hypervitaminosis.

Vitamin A has been synthesized in several different laboratories. There are really two forms of the vitamin. One form (A1) constitutes most of the vitamin content of the livers of salt water fishes. In the livers of fresh water fishes vitamin A_2 occurs. The 2 substances give slightly different colours with antimony trichloride and can be distinguished spectro-scopically.

In higher animals, vitamin A is important for growth. When rats are deprived of vitamin A, the cornea of the eye becomes horny. This condition is known as xerophthalmia. It can occur also in children. Not only the corneal epithelium, but other types of epithelia are also adversely affected in vitamin A deficiency; this may lead to lessened resistance to infection. Vitamin A has a relation to vision. In man, vitamin A deficiency causes night blindness, but the effect is not as readily produced as formerly supposed. In so far as they have been investigated, invertebrates, with the apparent exception of the snail Helix, do not seem to require vitamin A. Aside from its relation to vision, little is known as to the reason why vitamin A is necessary. Its presence in cells can be recognized with a fluorescence microscope because of the fact that in ultraviolet light it gives a green fluorescence.

STEROLS-VITAMIN D

Rickets, the primary vitamin D deficiency disease, is a skeletal disorder of young, growing animals generally characterized by decreased concentration of calcium and phosphorus in the organic matrices of cartilage and bone. Vitamin D deficiency results in clinical signs similar to those produced by deficiency of calcium and (or) phosphorus, because all three nutrients are required for proper bone formation. In the adult animal, osteomalacia is the counterpart of rickets. Cartilage growth in the adult has ceased, and thus this condition is characterized by a decreased concentration of calcium and phosphorus in the bone matrix (demineralization). Symptoms of rickets include the following skeletal changes, varying somewhat with species depending on anatomy and severity: (1) weakened long bones, resulting in curvature and deformation; (2) enlarged, painful hock and knee joints; (3) general stiffness of gait, arched back and a tendency to drag hind legs; and (4) beaded ribs and deformed thorax. With osteomalacia, in adults, the progressive demineralization of bones eventually results in fractures and breaks. The disturbance of calcium and phosphorus metabolism produces other symptoms such as reduced performance, hypocalcemia and reproductive failure.

A number of sterols have vitamin D activity, that is to say they prevent the occurrence of rickets in higher animals. The chemistry of the sterols is very complicated, and books on physiological chemistry may be consulted for structural formulae, etc. When the common sterol of fungi, ergosterol, is irradiated with ultraviolet rays, another sterol, calciferol, is produced. This is commonly referred to as vitamin D_2. (There is no D_1, for the original use of the term was a misnomer). D_3 is 7-dehydrocholesterol. Numerous other antirachitic (rickets-preventing) substances have also been described. Whether or not a given sterol or sterol derivative has antirachitic action can only be determined by actual test. Bacteria do not seem to require sterols. Some species of protozoan flagellates do; also Entamoeba histolytica, the ameba that causes human dysentery. Insects, in so far as they have been studied, require sterols, but they seem to need cholesterol, a cholesterol derivative, or a plant sterol; calciferol cannot usually be substituted for these. In mammals and birds, vitamin D causes an increase in the deposition of calcium and phosphorus in the bones.

This is apparently due, in part at least, to an increased absorption through the intestine. Using radioactive isotopes, Greenberg found that vitamin D promotes the absorption of calcium and strontium from the digestive tract. Sterols may act in the same way in promoting the passage of calcium through the walls of the insect intestine. Moreover, they may be concerned with the entrance of calcium into cells. This aspect of the subject has never been investigated.

VITAMIN E

Vitamin E is a the principle membrane-associated antioxidant molecule in mammals. It plays a major role in preventing oxidative damage to membrane lipids by scavenging free radicals. A third fat-soluble vitamin or group of vitamins is vitamin E, or perhaps one should say the vitamins E. There are very many substances with vitamin E activity, but the term vitamin E is restricted to the tocopherols. (The word tocopherol is derived from Greek words signifying child birth). There are 4 tocopherols now known: alpha, beta, gamma and delta tocopherol. The formula for alpha tocopherol is:

Alpha Tocopherol

Beta and gamma tocopherol have 1 less methyl group than alpha tocopherol and delta tocopherol has 2 less methyl groups. Vitamin E is necessary for the reproduction of the rat and other mammals and birds. If the male rat is fed a diet lacking in vitamin E, the germ cells in the testis degenerate. (The effect is not evident in the mouse, but is said to occur in the little fish commonly known as the guppy). In the absence of vitamin E, the female rat cannot give birth to its young; these degenerate and are resorbed. In some cases, muscular atrophy follows lack of vitamin E in the diet. In 1949, Harris wrote: "The mechanism whereby vitamin E functions in the body is still unknown."

A dual role was proposed by Hickman and Harris. In the first place it may act as part of an enzyme system, and to this action its effect in preventing degeneration of muscle may be due. Secondly, the tocopherols exert another, less specific function in that they tend to prevent excessive oxidation. Thus, for example, they tend to preserve vitamin A and they may also prevent the necessary unsaturated fats from being oxidized.

VITAMIN K

The major clinical sign of vitamin K deficiency in all species is impairment of blood coagulation (Griminger, 1984). Other clinical signs include low prothrombin levels, increased clotting time and hemorrhaging. In its most severe form, a lack of vitamin K will cause subcutaneous and internal hemorrhages, which can be fatal. Vitamin K deficiency can result from dietary deficiency, lack of microbial synthesis within the gut, inadequate intestinal absorption or inability of the liver to use the available vitamin K. Accidental dicumarol, or warfarin, poisoning should be considered in dogs or cats observed with hemorrhagic syndromes.

Fourth and last of the commonly recognized fat-soluble vitamins are the vitamins K, so named because they are important for blood coagulation (coagulation is spelled with a K in German and in the Scandinavian languages). The K vitamins are close to the E vitamins chemically. Tocopherols are very similar to the naphthoquinones chemically; the K vitamins are actually naphthoquinones. This is a 1,4 naphthoquinone. The most active form of vitamin K, a form which can be prepared synthetically, is 2-methyl-1,4-naphthoquinone. This is sometimes called menadione. It does not occur naturally nearly as abundantly as vitamin K1, which resembles menadione except that instead of having a methyl group in the 3 position, it has a phytol group.

(Phytol is a 20 carbon mono-unsaturated primary alcohol which can be obtained from the saponification of chlorophyll). Vitamin K_2 also resembles menadione except that in the 3 position it has a difarnesyl group. (Farnesol is similar to phytol; it is a 15 carbon primary alcohol with 3 unsaturated carbon atoms). Several other naturally occurring vitamin K compounds have also been described. They are all either yellow oils or yellow crystalline solids, practically insoluble in water. Vitamin K is manufactured in one form or another by bacteria, which do not ordinarily require it.

The bacteria in the intestines of mammals produce enough so that normally the mammals have no need of it from outside sources. However, when absorption of fatty compounds is interfered with, as in intestinal disease or in liver disease involving interruption of bile flow, the vitamin becomes essential for man and mammals. Many birds require the vitamin, for their bacteria can not supply them with sufficient quantity. Vitamin K is necessary for the clotting of the blood of higher animals. The reason for this is not clearly understood.

To some extent opinions differ, but most authorities believe that this action of vitamin K is an indirect effect caused by the entrance into the blood of greater amounts of a substance prothrombin, the precursor of thrombin. Thrombin, which is regarded as an enzyme, is of primary importance in blood clotting. Certainly the function of vitamin K can not be restricted to its effect on the clotting of blood. Obviously, the bacteria and the plants which produce it must have some need for it themselves. Experiments with isolated cells show that vitamin K is a very powerful protoplasmic clotting agent.

This is not surprising in view of the fact that the clotting of protoplasm is in so many respects similar to the clotting of blood. Although menadione is but sparingly soluble in water, solutions of it cause a coagulation of the protoplasm of marine eggs. Menadione also produces clotting in muscle protoplasm.

ASCORBIC ACID (VITAMIN C)

Vitamin C is important to all animals, including humans, because it is vital to the production of collagen. Vitamin C is also important because it helps protect the fat-soluble vitamins A and E as well as fatty acids from oxidation. Vitamin C prevents and cures the disease scurvy, and can be beneficial in the treatment of iron deficiency anemia.

The fat-soluble vitamins are still generally referred to by their alphabetical names. On the other hand, the water-soluble vitamins are now for the most part usually given chemical names which have no relation to the alphabet. This became necessary when what was originally the B vitamin became a whole assortment of heterogeneous substances, distinguished by somewhat variable and uncertain subscripts of the letter B. At present the water-soluble vitamins include the clearly defined ascorbic acid (vitamin C) plus the vitamins of the B complex. Because of the fact that vitamin C is one substance, clearly defined and understood, we shall consider it first rather than in its alphabetical order. Ascorbic acid is now well understood chemically and it can indeed be synthesized. The L-form is much more potent than the D-form. Ascorbic acid is really an acid hexose sugar. Its acidity is due to the dissociation of an enolic hydrogen rather than to opening of the lactone ring. Ascorbic acid is a strongly reducing substance, and because of this, it is easy to demonstrate its presence in cells and tissues. A standard method involves the reduction of silver nitrate. Because of its reducing power, there is doubtless a relation of ascorbic acid to the oxidative processes in cells, but there is still no clear understanding as to the exact function or functions of the vitamin in metabolism.

In animals suffering from scurvy there is lack of intercellular material; presumably the cells enclosing the blood vessels are not held together as firmly as they should be and this is doubtless the cause of hemorrhage. Mucopolysaccharides are known to be important constituents of the cementing substances which hold cells together. In view of the fact that mucopolysaccharides are polymers of acid sugars, it is possible that ascorbic acid is involved in the synthesis of such substances. There is some indirect evidence for this. Normally in wound healing mucopolysaccharides are detectable in the healing tissues and may have considerable importance for the process. However, when the store of ascorbic acid is depleted, these substances do not appear.

THIAMINE (ANEURINE, VITAMIN B1)

The formula for thiamine is $C_{12}H_{18}N_4OS$. It is prepared synthetically as the chloride hydrochloride. This is a white crystalline compound readily soluble in water. Thiamine is a combination of a pyrimidine ring and a sulfur-containing thiazole ring.

Apparently all sorts of protoplasm need thiamine. Some types of cells can synthesize it, others can not. Higher organisms generally require the intact thiamine molecule; on the other hand some lower organisms are able to synthesize thiamine provided they can obtain both parts of the molecule, pyrimidine and thiazole. Other organisms may require either the pyrimidine fraction or the thiazole fraction, but not both. Many bacteria require neither and can synthesize whatever thiamine they need.

Lack of thiamine in the diet produces beriberi, a paralytic disease of men rather common in Japan and in various other portions of the world. In birds, lack of thiamine produces polyneuritis, a disease similar to beriberi. In this disease the nerve fibres show degenerative changes. On a fox ranch in Minnesota owned by a Mr. Chastek, it was found that the inclusion of raw fish in the food given the foxes caused a paralysis now known as Chastek paralysis. Experimentally, it has been shown that a similar paralysis could be produced in chicks, cats, pigeons and rats. The effect is due to the fact that some species of fish contain in their tissues an enzyme, thiaminase, which destroys thiamine.

Thiaminase is found not only in fish but also in clams. In addition to its effect on the nervous system, lack of thiamine may also cause retardation of growth. Thiamine is important for the utilization of carbohydrate. Combined with 2 phosphate groups as diphosphothiamine, it forms part of an enzyme system that acts on pyruvic acid, a keto acid of great importance in intermediary metabolism. When animals are deprived of thiamine, their tissues are unable to oxidize pyruvic acid. According to von Muralt, thiamine is also important in the transmission of the nerve impulse. When isolated nerves are stimulated, they are believed to release thiamine to the surrounding solution.

RIBOFLAVIN

Riboflavin is an orange crystalline substance, sparingly soluble in water, and markedly sensitive to light. The riboflavin molecule is complex. For an understanding of its structure and its synthesis as first accomplished by Karrer and his associates in 1935, the larger textbooks of organic chemistry may be consulted. The three rings represent an isoalloxazine constituent and this is combined with ribose, the pentose sugar also present in ribonucleic acids. Although riboflavin is apparently a constituent of all or almost all types of protoplasm, many lower organisms do not require it in their food.

Yeasts and molds have no need for it. Bacteria, in general, can get along without it. However, riboflavin serves as a growth factor for the tetanus bacillus, for lactic acid and propionic acid bacteria, and it is a requirement for Erysipelothrix rhusiopathiae and Listerella monocytogenes. The protozoan ciliates that have been investigated require riboflavin. So, too, do almost all the insects that have been studied. Riboflavin is essential for the growth and normal health of the rat, mouse, dog, pig, man, chicken and turkey. In riboflavin deficiency, various symptoms of ill health may appear.

These include redness and roughness of the cornea of the eye, soreness at the corners of the lips, redness and soreness of the tongue, etc. At one time, it was thought that "ariboflavinosis," the disease caused by riboflavin deficiency, was extremely common. But the symptoms mentioned above are not always due to lack of riboflavin. The presence of riboflavin in cells may be detected by fluorescence microscopy. Riboflavin is important for cell respiration. It and other alloxazine compounds combine with phosphoric acid and protein to form enzyme systems known as flavoproteins.

NICOTINIC ACID (NIACIN)

Nicotinic acid and pyridoxine (to be considered next) are closely related substances. Both are derivatives of pyridine. This substance is a weak base, the molecule of which contains five CH-groups and a nitrogen atom arranged in a six-membered ring. As is customary with ring compounds, the various positions on the ring have numbers. In pyridine, nitrogen occupies the number 1 position. Nicotinic acid is 3-pyridine carboxylic acid. The structural formulas of the acid and its amide are given below:

The amide is more important than the acid itself, for some organisms require it and cannot manufacture it from the acid. Both acid and amide are white crystalline powders, without odour. According to Stephenson, "Nicotinic acid is one of the commonest growth factors needed by bacteria; in other words the ability to synthesize it is one most frequently lost. This may be due to its very wide distribution in soil and natural media." It is an essential requirement for those ciliate protozoa that have been studied (Colpoda and Tetrahymena) and also for most of the insects that have been investigated. Nicotinic acid is essential for the dog, the pig, and for man. However, if enough tryptophan is added to the diet, neither dog nor pig requites it.

As noted previously, tryptophan is a precursor of nicotinic acid, and the tissues can synthesize the vitamin from tryptophan. From a lack of nicotinic acid, dogs become afflicted with what is known as blacktongue disease. In man, nicotinic acid deficiency is the primary cause of pellagra. This serious disease, common enough in the United States, involves the lining of the digestive tract, the skin, and the nervous system. Nicotinic acid amide enters into the composition of two substances of great importance for the oxidative metabolism of the cell.

These are coenzyme I and coenzyme II. Presumably nicotinic acid deficiency is associated with a lack in the tissues of coenzymes I and II. But what connection there is between such a lack and the manifold disturbances present in a disease like pellagra is not at all clear.

PYRIDOXINE (VITAMIN B6)

The term vitamin B6 is now properly used, according to a decision of the American Society of Biological Chemists, to designate three naturally occurring substances: pyridoxine, pyridoxal and pyridoxamine. Pyridoxal is the aldehyde derivative of pyridoxine and has a formyl group instead of a hydroxymethyl group at the 4 position. Pyridoxamine is an amino derivative and has an amino methyl instead of a hydroxymethyl group at the 4 position. The three forms of the vitamin are almost equally effective for higher animals, but some bacteria have specific requirements for one form or the other. Vitamin B_6 is needed by many bacteria, by some protozoa, and by most of the insects whose vitamin requirements have been investigated.

When rats are deprived of vitamin B_6, they develop a painful inflammation of the skin. This is especially evident in the skin of the extremities—the paws, snout, and ears. The ailment is sometimes called rat acrodynia because of a supposed similarity to a human ailment in which the nerves and skin of the hands and feet are affected.

The disease is not wholly specific for vitamin B_6 deficiency; it can be produced also in other ways, for example by a lack of unsaturated fatty acids. In various higher animals, convulsions may follow extreme vitamin B_6 deficiency. Sometimes lack of the vitamin causes muscular weakness; it can also cause anemia. Vitamin B_6 deficiency can be induced by the addition of an antagonistic analogue, desoxypyridoxine. Thus if this compound is injected into hens' eggs, the embryos do not proceed very far in their development. Vitamin B_6 is important for the proper metabolism of the amino acids. It forms part of an enzyme system which promotes the synthesis of amino acids by the addition of carboxyl groups. Acting as part of what is presumably the same enzyme system, the vitamin also favours the removal of carboxyl groups from amino acids.

Another type of enzyme for which vitamin B_6 is an important constituent is transaminase. This enzyme catalyzes the transfer of an amino group from one amino acid to another; an important type of reaction in protein metabolism. Thus there is some important information as to how the vitamin functions with respect to the chemical reactions occurring during the metabolism of proteins and amino acids, but how this is related to the disturbances which occur in skin, nerves, muscle and blood when the vitamin is lacking has yet to be explained. It has been claimed, on the basis of isolation studies of mitochondria, that these granules contain especially large amounts of vitamin B_6. However, it is the nucleoprotein-containing granules of the cytoplasm that are believed to be especially important for protein metabolism.

BIOTIN

As long ago as 1898 it was shown that if dogs were fed raw egg white, diarrhea resulted. Following this it was shown that other mammals also suffered from egg white feeding, and that in addition to the diarrhea, skin ailments and nervous symptoms appeared. The phenomenon became known as egg white injury. Cooking the egg white prevented the injury. György proposed the term vitamin H for a factor which prevented egg white injury. Later, it was shown that this factor is identical with biotin, a vitamin known to be important for yeasts and microorganisms. The raw egg white unites with the biotin and prevents its utilization by mammals or birds. Chemically, biotin is rather complicated. It has in its molecule a urea ring containing one sulfur atom; the ring is combined with a valeric acid group. The presence of sulfur in the molecule is not important for the vitamin activity, for if the sulfur is replaced by oxygen, the vitamin activity is retained. Biotin acts in extremely low concentration, and it is thus one of the most potent of the vitamins. Many organisms do not grow unless biotin is available for them. Usually in higher organisms it is synthesized in sufficient quantity, but biotin deficiency can occur naturally in birds and in insects.

Lack of sufficient biotin has the same effect as egg white injury—diarrhea, disturbances in the skin and nervous system, in rats loss of hair about the eyes. It can also prevent the birth of young in rats. The protein in egg white which binds biotin is called avidin; it is a basic, carbohydrate-containing protein and it occurs not only in the eggs of birds but also in frog eggs. There is now strong evidence that biotin is important in a process which is known as carbon dioxide fixation. Many bacteria, and higher organisms also, are able to utilize carbon dioxide and form organic compounds from it. This process is interfered with if biotin is lacking.

PANTOTHENIC ACID

Pantothenic acid is found in many diverse types of living materials. Chemically, it consists of β-alanine in a peptid linkage with a dihydroxy acid. (β-alanine differs from ordinary alanine in being β-aminopropionic acid instead of α-aminopropionic acid). The acid is unstable and for this reason is sold as a calcium salt. This salt is a white powder, readily soluble in water but not in alcohol. Many organisms from bacteria and protozoa to higher mammals require pantothenic acid. In its absence these organisms do not grow properly and they may show various ill effects.

Symptoms of pantothenic acid deficiency include dermatitis (inflammation of the skin) in rats, deficient feathering in chicks, graying of hair in rats, degeneration of the nervous system in pigs, hemorrhages in the adrenal gland and general disturbance of function of this gland, etc. The vitamin constitutes part of an enzyme system. Combined with various substances including phosphate, adenylic acid, and a sulfhydryl group, it is coenzyme A, and this in combination with a protein is an enzyme capable of acetylating various organic substances such as choline and sulfanilamide. The acetylation of choline may be very important physiologically. How the acetylation, or rather the lack of it, can produce all the pathological conditions noted above is not clear, although in the case of the nervous system, any alteration of acetylcholine metabolism would certainly be harmful.

FOLIC ACID

This vitamin occurs in green leaves; hence the name, which is taken from the Latin word for leaf. Folic acid also occurs in mushrooms and yeast; and in animal tissues, such as liver and kidney. The vitamin has been called by many different names, and it exists in various forms. These have different potencies for different types of organisms. Chemically, folic acid consists of pteroic acid combined with the common amino acid, glutamic acid. Pteroic acid itself is a combination of para-aminobenzoic acid and a pteridine base. Pteridine bases are pyrimidine derivatives related to the pyrimidine bases thymine and uracil found in the nucleic acids.

The compound shown above is pteroylglutamic acid, sometimes called PGA. Pteroic acid may also combine with 3 glutamic acid groups to form pteroyltriglutarnic acid (P3GA) or with 7 molecules of glutamic acid to form pteroylh-eptaglutamic acid (P7GA). All of these compounds are found naturally and all have vitamin activity for certain types of living materials. Folic acid is required by some bacteria, by certain ciliate protozoal, by some insects and by various birds and mammals. The bacteria in the rat intestine normally manufacture a sufficient quantity of the vitamin, and treatment with antibiotics such as sulfa drugs is necessary to cause folic acid deficiency.

Folic acid is commonly an essential growth factor, not only for bacteria but for higher organisms as well. Lack of the vitamin in mammals results in various ailments, but especially in a type of anemia characterized by a small number of red blood cells of large size. This type of anemia, called macrocytic anemia, occurs in the tropical disease sprue, and sprue can be cured either by liver extract (which contains folic acid) or by folic acid itself. Folic acid is apparently important both for growth and cell division. When chick embryos are in contact with antagonistic analogues such as 4amino folic acid, their growth is greatly retarded. Similar results have been obtained in numerous other studies with folic acid antagonists. Thus 4-amino folic acid has a retarding effect on tumor growth. A substance related to folic acid is the so-called "citrovorum factor," which is apparently identical with "folinic acid." This factor favours growth of the bacterium, Leuconostoc citrovorum. The citrovorum factor may be obtained from liver extracts; also it is excreted into the urine of rats or men fed folic acid.

OTHER POSSIBLE VITAMINS OF THE B COMPLEX

B_{13} has been described as a growth factor for rats and for pigs. There is also a description of a vitamin B_{14}, extracted from urine. Over the years, there has been a large and controversial literature about what has been called vitamin P. Substances with vitamin P activity are supposed to maintain proper permeability of the blood capillaries of higher animals and to prevent excessive loss of fluid from them. (The letter P is due to this permeability action). There is now general agreement that substances with P activity are not to be considered as vitamins. *Para-aminobenzoic acid* (often called paba). This relatively simple compound is an important growth factor for many bacteria. Thus sulfanilamide is an antagonistic analogue of para-aminobenzoic acid, and its antibacterial action is due to this similarity in structure. The discovery of this relationship in 1940 was indeed a milestone in our understanding of the mechanism of drug action and has served as a model and guide for many subsequent researches.

Choline

This substance, the basic constituent of lecithin, is believed by many to be a vitamin. Rats deprived of choline develop fatty livers and other types of hepatic injury. They may also show degenerative changes in the kidney. Choline deficiency is also a factor in the development of a leg deficiency (perosis) in chicks and turkeys.

It should be remembered that choline is a constituent not only of lecithin, but also of the important humoural substance, acetylcholine. For a discussion of the significance of acetylcholine, *Inositol*. Inositol is a derivative of benzene. It can be prepared from hexahydroxy-benzene by reduction, in the course of which 6 hydrogen atoms are added.

Thus inositol is hexahydroxy-hexahydro-benzene. It is an optically active substance found in yeast, muscle, and in various other types of animal and plant tissue. Inositol is important for the growth of yeasts, and is indispensable for some types of yeast.

It also promotes the growth of some fungi. In the presence of inositol, chicks grow more rapidly than when it is absent. However, rats may be bred for three generations without inositol, and György was led to conclude that "the facts presented do not, at the present time, warrant the identification of inositol as a primary and really essential vitamin; they rather favour the assumption that it is a supporting and at least not always specific vitamin." *α-Lipoic Acid*. This substance, also called protogen, is probably to be included among the vitamins of the B group, for it is water soluble.

It is important for the growth of some bacteria and for the protozoan Tetrahymena, and it is involved in the oxidation of the important keto acid, pyruvic acid (for this reason it was formerly called pyruvate oxidase factor). Many types of living material contain lipoic acid. It is now known to be a derivative of octanoic (caprylic) acid and it contains sulfur in the form of an S-S group.

There are doubtless other vitamins of the B group still to be discovered. Fraenkel found that the larva of the beetle Tenebrio molitor (the so-called mealworm) requires some substance present in yeast or liver but not one of the known B vitamins. This he called B_T. More recently he has identified it with carnitine, a dipeptid known originally in muscle.

Also the flour beetle requires at least two other unidentified members of the B complex and neither of these is B_T. Goetsch claims that there is a vitamin present in certain insects, especially termites, a vitamin which has pronounced effects on growth and differentiation. This he calls the vitamin T complex; it is supposed to affect not only insects but various other types of organisms. Throughout the course of vitamin study, there have indeed been many extravagant claims and many false leads. The vast literature is often contradictory and confused. But in spite of this, in the course of time the truth has gradually emerged and the main outlines of the picture are now clear.

All sorts of organisms, from bacteria to man, require certain specific types of organic molecules. These can be synthesized by green plants and by many lower organisms. Higher forms have frequently lost the power for such synthesis, and unless, like the ruminants, they provide themselves with huge storehouses for bacteria, they must depend for their existence on obtaining the vitamins with their food. From the knowledge of the nature of the vitamins, we can begin to develop a chemical anatomy of living systems. The vitamins represent indispensable organic molecules and apparently the same molecules are almost universally needed. Moreover, we are beginning to discover why the vitamins are needed and what work they do in the living cell. Much of our information is still fragmentary.

The vitamin requirements for the lower vertebrates and for the hosts of invertebrate animals have scarcely been investigated at all. Moreover, in some instances we are completely at a loss to account for the fundamental necessity for a given vitamin. Such superficial disturbances as a reddening of the eye or a scaliness of the skin give no direct indication of the part a vitamin may play in protoplasmic activity.

VITAMIN B12

Year by year, the number of B vitamins increases. B_{10} and B_{11} are required by chicks, the former for proper feathering, the latter for growth. From liver it has been possible to isolate a pink crystalline substance which can be used to cure pernicious anemia. Folic acid is useful in this disease, for it has a hematopoietic effect (that is to say, it favours increased production of blood cells), but it does not correct the nervous symptoms of the disease. These depend on degenerative changes in the nervous system. Vitamin B_{12} in tiny amounts has been thought to correct diseases of both blood and nervous systems, but apparently if the nervous system is badly affected, it can not be cured.

The vitamin has been crystallized. It contains 4-4.5 per cent cobalt and also phosphorus; the molecular weight is approximately 1500. (Estimates vary from 1300 to 1600). In liver extracts there are really two active substances with slightly different absorption spectra. One of these is the definitive B_{12}, the other is B_{12b}. (B_{12a} is another similar substance derived from B_{12} by hydrogenation). Vitamin B_{12} is apparently identical with what has been called "animal protein factor." This was originally obtained from such diverse sources as milk, muscle, liver and cow manure. Although called animal protein factor, it can likewise be obtained from plant sources. Both animal protein factor and vitamin B_{12} promote growth in some bacteria, in chickens, rats and mice. The effect on growth may be due to the fact that the vitamin seems to increase the utilization of amino acids. At any rate, this has been reported for chicks.

Index